WITHDRAWN

JAN 1 2 2011

S0-BAM-685

The Maternal Factor

The Maternal Factor

Two Paths to Morality

Nel Noddings

UNIVERSITY OF CALIFORNIA PRESS

Berkeley   Los Angeles   London

University of California Press, one of the most
distinguished university presses in the United States,
enriches lives around the world by advancing scholarship
in the humanities, social sciences, and natural sciences. Its
activities are supported by the UC Press Foundation and
by philanthropic contributions from individuals and
institutions. For more information, visit www.ucpress.edu.

University of California Press
Berkeley and Los Angeles, California

University of California Press, Ltd.
London, England

© 2010 by The Regents of the University of California

Library of Congress Cataloging-in-Publication Data

Noddings, Nel.
    The maternal factor : two paths to morality /
Nel Noddings.
        p.   cm.
    Includes bibliographical references (p.   ) and index.
    ISBN 978-0-520-26549-3 (cloth : alk. paper)
    ISBN 978-0-520-26550-9 (pbk. : alk. paper)
    1. Caring.   2. Motherhood—Moral and ethical
aspects.   3. Ethics, Evolutionary.   4. Feminist ethics.
I. Title.
    BJ1475.N625   2010
    171'.7—dc22   2009034344

Manufactured in the United States of America

19   18   17   16   15   14   13   12   11   10

10   9   8   7   6   5   4   3   2   1

This book is printed on Cascades Enviro 100, a 100% post
consumer waste, recycled, de-inked fiber. FSC recycled
certified and processed chlorine free. It is acid free,
Ecologo certified, and manufactured by BioGas energy.

CONTENTS

ACKNOWLEDGMENTS

Several people have contributed generously to this work by reading drafts, discussing problems, and suggesting further readings. Virginia Held and Sara Ruddick were interested and enormously helpful while the book was little more than a gleam in my eye, and I have drawn on their work in addition to their valuable feedback.

Harvey Siegel, Marc Bekoff, Jean Watson, and David Glidden provided useful suggestions and much-appreciated encouragement. Conversations with Lynda Stone, Lawrence Blum, Nona Lyons, and Steve Thornton aided my thinking.

I am also indebted to audiences who responded enthusiastically to my initial ideas, among them those at Penn State, Lewis and Clark, University of Hawaii, Seton Hall, Miami University of Ohio, St. Benedict's, the Holistic Education Association, and the Mountain Lake Music Colloquium.

Michael Slote, as both formal and informal reviewer, saved me from a couple of egregious historical errors and, more important, asked the hard questions that had to be answered in the final manuscript. Many, many thanks.

As always, my husband, Jim, deserves thanks (and perhaps a medal) for putting up with my tossing and turning, grumbling, and moments of perhaps unwarranted enthusiasm.

Finally, I want to thank Naomi Schneider, patient and resourceful editor; Marilyn Schwartz, managing editor with a talent for diplomacy; and Mary Ray Worley, copyeditor who did an exemplary job.

INTRODUCTION

Much work is being done today on the evolution of morality. Anthropologists, psychologists, evolution scientists, and philosophers are looking for the roots of altruism, empathy, solidarity, and cooperation.[1] Surprisingly, in seeking these roots, scholars rarely look at female experience. It may well be that one wide and increasingly influential approach to moral life—care ethics—can be traced to maternal instinct.

I will not argue that an ethic of care evolves in a blindly biological way. Thinking, experimenting, reflecting, analyzing, and conceptualizing are all involved in developing an ethic. As Virginia Held has argued, when we consider naturalizing morality, we should not lose sight of ourselves as moral subjects—as persons who think and make choices, some of which challenge the way we are as biological creatures.[2] Cognition and the power to evaluate have also evolved, and we have the ability to reflect on our own nature. The ethic of care can be described as naturalistic in its rejection of supernaturalism and its recognition that science can study moral behavior and experience. It is also naturalistic in its insistence on remaining closely aware of actual human conditions and human nature. But it is not narrowly naturalistic or reductionist. There are elements of human emotion, attachments, and choice that demand a normative orientation, one that may challenge our natural inclinations; care ethics recognizes a moral subject.

We can ask, however, how this moral subject arises from natural beginnings, and we can examine these natural beginnings to locate problems that may beset a normative ethic under construction.[3] For example, the current study of altruism has revealed that altruism most often occurs between entities closely related by bloodlines. But we also know that, in the absence of hostile relations, many human beings respond sympathetically to the pain and needs of strangers. If it is true that we are naturally more inclined to care for our own children than for those of others, how do we prod ourselves to overcome what might be a moral shortcoming, and how do we decide that it *is* a shortcoming?

In searching female experience for the roots of morality, we should be astonished if thousands of years of confinement to home, family, and small community did not produce evolutionary effects. Among the salutary effects are female sensitivities to the needs of others and inclinations to respond positively to those needs. Among the negative effects may be an acceptance of subordination and a lack of facility with objects, spatial relations, and abstractions connected with the realm of human-made objects. We'll have much to say about the acceptance of subordination in what follows. Even to suggest the possibility of the second set of evolutionary effects is to risk condemnation. But to suggest the possibility of positive evolutionary effects, one must also be willing to at least entertain the possibility of negative effects. We must proceed with caution and open minds.

On the positive side, we should consider surveys that show, year after year, that women are, in general, significantly more concerned with social issues than are men. When we say "in general," we mean that, if the studies are well done, we may expect comparable results from further studies—that is, that more women than men will declare interest in matters such as child welfare, poverty, gun control, care of the aged, and compassionate treatment of prisoners. Such generalizations tell us nothing about the next woman or man whom we may meet. We cannot extrapolate from such generalizations to individuals. Similarly, on the negative side, we know that females do not do as well as males on mathematics tests and like measurements of ability in science and engineering. People dispute this conclusion by pointing to the fact that more girls than boys are now taking advanced courses in high school mathematics and getting better grades than boys. More girls also go on to college. But girls still score about thirty-five points lower than boys do on the math SAT. Similarly, girls rarely rank at the highest levels in science competitions, although that picture is changing. In 2007, for the first time, girls won top honors in the Siemens Competition in Math, Science, and Technology.[4] This result may vindicate those who insist that female/male differences in math aptitude are entirely a result of socialization and educational opportunities. Those who make this claim may be right. Unfortunately, the results are complicated by the fact that three-quarters of the finalists in the competition have a parent who is a scientist, so there are almost certainly both genetic and socialization factors operating. In this case, parental genes may have trumped the genes that might predict female/male differences. One could argue, perhaps even more persuasively, that the outstanding female students had the benefit of homes and parents who are scientifically well educated.

It would be quite wonderful if what has long seemed an evolutionary fact turns out to be a mere cultural artifact. The twenty-first century might see a fair share of female mathematical geniuses. But readers will become aware, as the book proceeds, that my interest lies more in the possibility of changing a male evolutionary trait—the tendency to aggression and violence. If we celebrate the success of women becoming more like men in professional life, can we at least speculate on the possibility of men becoming more like women in peacemaking, tenderness, and nurturance?

This book is not, then, about gender differences in math and science. Such differences are mentioned in part to remind readers that we should keep our minds open to possibilities that we do not welcome as well as to those we do. I use the current attempt to close the "gender gap" in math and science as a contrast to a gender gap that is far more important for the future of human life—that between male aggression and female nonviolence. We should also be aware that these are complex matters, and human performance is influenced by many factors other than genetics. When a difference is recognized, we can deny that it has any genetic source at all; we can even deny that the difference exists. That is the popular course of action taken today, often with the best intentions. Alternatively, we might ask whether we should try to reduce the difference. Is the difference a product only of environment? It might be. Then what changes should be made? But we should also ask why we value (or fail to value) the attribute or activity in which one group exceeds the other. Why, for example, is mathematical talent valued more than the talents and sensitivity required to meet the needs of children? Why are the traits traditionally associated with women undervalued?

Current and future advances in evolutionary studies and genetics might produce other results we would prefer not to consider—results on racial/cultural differences in abstract intelligence, for example. Most of us hope and believe that such differences do not exist. But we should be prepared for a variety of possibilities. Instead of steadfastly insisting that no such differences exist, we might question the scale of human values we have constructed. Why, for example, should abstract/mathematical talent be rated so much more highly than other human talents? That is a good question to ask even if genetic studies bear out our best hopes. It will be a significant question in our examination of caregiving.

Care ethics can provide guidance on how to respond to differences that are already obvious and to those that may emerge in further genetic studies. As carers, we are attentive to expressed needs and wants, and we are committed to respond to them. We are not governed entirely by assumed needs—those we establish tentatively before even meeting partners in dialogue—children, students, patients, clients. In teaching, we do not decide how to guide or instruct students solely on the basis of what we "know" about various groups. When girls want to study math or science, we should give them unreserved support. But when a bright girl expresses her desire to teach young children, we should also support that decision; we should *not* tell her, as so many advisers do today, "You're too smart for that."

In the first chapters of this book, I will try to make a plausible case for the beginnings of a distinctive moral approach arising from maternal instinct. Capacities such as the ability to "read" the infant and its needs develop. A similar capacity to read male partners who may or may not act as protector for both mother and infant emerges. Females who did not develop the capacity for "empathy" (the word is problematic and will need analysis) might not have survived and, of course, neither would their infants.[5]

But how might a moral subject or agent emerge from these instinctive beginnings? How might women have become concerned about the well-being of other children and other mothers? What concepts and commitments are involved in the transition from instinctive caring to moral caring? What special problems arise? (Readers should understand that I am not claiming that the path from maternal instinct is the only path to full moral life, but it is one primary candidate that has been neglected.)

From the start, mother and infant are bound in a relation that, if the infant is to survive and thrive, must be a caring relation. The female's relation with the male protector has, historically, been more instrumental than affectionate but sometimes develops into one of caring. However, that development is colored by the female's acceptance of a subservient position. The necessary acceptance of subservience might well increase the female's empathic skills, but it also leaves her vulnerable to all forms of apparently legitimate patriarchal power. For example, although the ethic of care has no need for recourse to the supernatural, women have been peculiarly susceptible to religious belief and practice. They have indeed worked hard to sustain a constellation of institutions that have denigrated their qualities and oppressed them. A well-developed ethic of care should help women to free themselves from this oppressive tradition.

The first three chapters will lay out the basic ideas of caring as an approach to moral life. Special consideration will be given to the caring relation and to the state of consciousness of the one caring: the receptive attention that tries to discern "what is there" in the recipient of care and motivational displacement— the flow of motive energy from the carer's personal objectives to those of the cared-for. In the past twenty-five years, much has been written about these features of caring, and it is necessary to answer some vital questions that have arisen. It is especially important to elaborate on the idea of *natural caring* which has triggered much misunderstanding. What is natural caring? How does it differ from ethical caring? And why might we argue that natural caring should be preferred to ethical caring?

The discussion of relatedness and the ontological status of relation leads logically to questions about autonomy. In chapter 4, I explore problems and promises associated with the concept of autonomy. Is there such a thing in real life as autonomy? How should it be described? Is there a view of autonomy compatible with care theory? What features usually associated with autonomy should be examined skeptically?

It has sometimes been suggested that care ethics should be considered a form of virtue ethics. In chapter 5, I argue that, although the two have obvious similarities, they are distinctly different, and I'll discuss differences between care ethics and virtue ethics, Confucianism, and Christian agape. Because care ethics—like the three views with which it might be associated—gives but a small role to principles in moral theory, might it provide a foundation for a form of feminist communitarianism? Why might we want to avoid this?

Several proponents of care ethics (and some critics too) trace its roots to moral sentimentalism—the idea that emotion, not reason, provides the motivation for morality. In chapter 6, a debt is acknowledged to the moral "sentimentalists"—especially Hume—but, again, we will see significant differences. Some of these differences might be regarded as improvements on the initial ideas enabled by information now available to us through evolutionary biology. Other differences are revealed by looking more appreciatively at contrasts between female and male emotions and how these differences encourage an alternative account of social virtues.

In chapter 7, the focus is on needs as a fundamental concern of care ethics. In caring, a carer tries to identify and, often, to meet the needs of the cared-for. Working with needs involves several conceptual difficulties. For example, it seems relatively easy to identify needs and negotiate among them in the context of families and small communities, but it is notoriously difficult in larger settings. Must we switch from care ethics to justice when we move from family circles to national and international settings? Is there promise in the notion of a care-driven conception of justice? How do female/male differences in experience (and thus in evolutionary tendencies) influence the way we approach the identification and satisfaction of needs?

Perhaps the least desirable legacy of human evolution is the lingering male tendency to aggression and domination. Chapter 8 examines war and violence and what care theory might contribute to their elimination or reduction. It is not an optimistic chapter. The male evolutionary tendency to aggression and domination has been aggravated by a pattern of cultural evolution that has applauded aggression and identified the desire to dominate with *manliness*. Education might work to transform this powerful and damaging model of masculinity, but such a transformation will be strongly resisted. Nevertheless, I offer some suggestions.

The final chapter offers a summary of main points and some speculation on the possibilities for a convergence of care theory and traditional male ethics. It should be obvious that no convergence can occur if there are not two or more identifiable streams of thought to converge. In the past two decades, care theory has developed in strength and popularity, but it is still too often thought to be just a branch of feminist ethics. The object of continued analysis and argumentation is to establish care ethics (or to show that it has been established) as a major alternative to traditional moral theories. Then true convergence will become a real possibility.

ONE

The Evolution of Morality

In this book, I am interested primarily in the evolution of morality through female experience and how that morality might be described. It makes sense, then, to start with a discussion of maternal instinct, infant bonding, and the empathic capacities developed through the basic experience of mothering. After laying out this story, we'll look at some current work on the evolution of morality—work that often ignores female experience entirely. The chapter will conclude with an outline of topics and questions to be addressed in later chapters.

A LIKELY STORY

The instinct to nurture and care for offspring is basic to the survival of every species. Some of the "lower" animals—insects, fish, and reptiles—may simply prepare the natal environment so that the newly hatched are likely to survive without the attention of a parent. But most mammals must provide food and protection for their babies until they are ready to fend for themselves. It is usually the female who cares for the young. Sometimes a male provides protection for both mother and infants, but some males have been known to threaten or even destroy the young. Then the mother, with help from one male or other females, must defend her babies. Often she will do this at the risk of her own life.

It is hard even to imagine what the earliest human females must have suffered in birthing and rearing their children. The best we can do is to tell a plausible story—some of it based on what we know about our likely ancestors among the primates. But certainly, a successful female—one who survived and nurtured more than one living infant—was aided by capacities associated with the maternal instinct. Maternal instinct prodded her to care about her infant's survival. She wanted it to live. To care successfully for it, however, she had to learn how "to read" the infant, understand the needs expressed, and have the resources to meet these needs. Is the infant hungry? Cold? In pain? In need of grooming? Females incapable of this basic reading would be unlikely to have surviving infants. Each new generation of females would likely have a propensity for what might be called elementary empathy.

It is necessary now to interrupt our likely story to say more about this crucial word *empathy*. With Martin Hoffman and Michael Slote, I want empathy to include both cognitive and emotional processes that help us to understand and sympathize with what another is experiencing.[1] Hoffman says: "The key requirement of an empathic response according to my definition is *the involvement of psychological processes that make a person have feelings that are more congruent with another's situation than with his own situation."*[2]

Donald Broom also refers to empathy as a process: *"Empathy is the process of understanding the experience of another individual, cognitively and emotionally."*[3] But this definition, too, begs for a description of related—perhaps necessary—processes under discussion here, such as receptive attention and the "reading" (cognitive apprehension) of others.

The designation "psychological processes" covers a large territory. What psychological processes are involved? Given this definition, how is empathy different from sympathy? Historically, "feeling with" or "feeling for" has been identified with sympathy, and the original use of *empathy* (first in aesthetics and then in psychology) referred to a cognitive process of understanding.[4] It did not mean taking on the other's feelings but, rather, understanding the other's perspective. As such, empathy could be used for instrumental or immoral purposes as well as moral ones.

In earlier work, I expressed some reluctance to use the word *empathy* because of these ambiguities.[5] But it represents a powerful shorthand, and I will use it occasionally here, with the understanding that it has both cognitive and emotional elements, and that the emotional element is primary. In care ethics, empathy is regarded as receptive, not projective. We do not "project" ourselves into the other in order to understand. In chapter 2, the processes involved in empathy will be examined more closely.

Now, back to our story. The earliest human mothers had to "read" their infants and respond to their expressed needs. Clearly, feeling is involved in this early empathy. The mother felt something as a result of her infant's expression of need. But then she had to assess the need; this is the "reading" process to which I've been referring. I will call needs so discerned *expressed* needs, because the mother's intention is to discover what the infant feels as a need. In contrast, a mother might assume a need without considering the child's expression. In much of what will be developed later, the distinction between *expressed* needs and *assumed* needs will be important.

Having decided what the child's expression of need means, the mother must respond to meet the need. In this "I must," we see a pre-moral imperative. The mother is not obeying some moral principle; she is responding quite naturally to the child's need, for the child's sake. But again, thinking is required in deciding what to do.

The problems involved in finding and using resources had to be pressing and omnipresent in the earliest days of human life on earth. Although we do not know details of early mating practices, we do know that the first humans, like their primate ancestors, were social beings. Why did females select and settle down with one mate instead of raising their offspring in enclaves of females? This choice is still veiled in mystery. Perhaps permanent affiliation with one strong male gave more protection for a woman and her babies. This seems reasonable. The female, naturally bonded with her infant, would seek or accept a male who would defend her child. Males might accept such responsibility in return for the ready availability of sex and the assurance that resulting offspring were his own. Having a dependable, readily available source for sex might well have been seen as more convenient than fighting other males for every opportunity. If, further, the male could secure more than one female, the arrangement might be very attractive to both male and females, who would form a social group to provide further protection for the young. The female's main criterion for selection of the male would be his strength and willingness to protect her and her young.

The male, however, was rarely as committed to the young as the female. Although we now have evidence that concern—even altruism—shows up in nonhuman species along bloodlines, paternity (until recently) has been an inference;[6] it is not so easily established as maternity. It is almost certain that the female had to keep the male satisfied if she wished to retain his services as protector and provider of some resources. Thus, in addition to learning to read her infant, she also had to read her mate. Was he hungry? Angry? Wanting sex? Wanting to be left alone?

While both relationships contributed to the female's evolving capacity for empathy, the one with her mate also conditioned her to subservience and dependency. Unable to protect her young alone and unable to achieve her ends by force, she accepted a role complicit in her own subordination. Care theory urges an end to this subordination. Some feminists worry that the ethic of care will actually contribute to the continued subordination and exploitation of women, but I will show that such an assessment rests on a misunderstanding.

Notice that as we seek the evolutionary sources of morality, we are looking at social beings from the start. Both males and females start out in relation, and females are virtually defined in relation. Frans de Waal, whose work we will consider again in chapter 6, acknowledges that women understand connection and relation better than men do, but he insists that connectedness applies also to men.[7]

But how does connectedness apply to men, and why have they so often rejected the idea in favor of individualism? Why have some men felt the need to construct elaborate theories to explain and defend social connectedness? Somewhere along the way, males did become more invested in the lives of their offspring. That investment may have been prompted, as already suggested, by the desire for dependable sex and other creature comforts. But we can be quite sure that a male's intense interest in establishing conditions to ensure that the offspring were indeed *his* is at least in part a reflection of the tendency humans share with the nonhuman world to identify with members of a kin-group.[8] In historical times, male interest in offspring also became bound up with interest in property. The offspring as well as their mother were considered the male's property. Further, even today, although both mothers and fathers sometimes abandon their children, it is far more common for fathers to do so. And, when children are killed by a parent, it is much more likely that the murder was committed by the father. The murder of a child by its mother occupies public attention for months, even years.

As surviving females grow in the capacity for empathy, their offspring—both female and male (to a lesser extent)—may also develop the capacity. The mother-child relationship begins before birth. We know now that a fetus in the last trimester of pregnancy can distinguish the mother's voice from that of others, and this may indicate a rudimentary capacity for empathy.[9] However, genetic differences in empathic response between males and females are probably there from the start, and subsequent socialization strengthens the difference.[10]

The maternal instinct in females is accompanied by biological responses that encourage empathy. For example, a crying infant—even one unrelated to the mother—will induce a letting down of milk in a lactating female and a tingling in the breasts of those who are not lactating. This biological response may well be accompanied by the customary feelings of sympathy and urgency a woman has for her own child, and it may provide a basis for the development of natural caring beyond maternal instinct. Similarly, females likely developed concern for other females who were nourishing their babies. They were aware of what other mothers were feeling—another extension of empathy.

In the relation between mother and child, we detect also the pre-moral "I must" or "I ought." No matter the mother's own activities, exhaustion, or fear, the imperative "I must" arises in connection with satisfying the infant's needs. As in the case of the lactating mother and a child not her own, the "I must" may arise in response to the needs of some others. When this response is generalized to wider situations, it becomes a genuine moral approach to life, and it is this beginning on which we may build an ethic of care. Before tackling the task of articulating such an ethic, we should engage in some discussion about the meaning of morality and what we are looking for in the present exploration.

MORALITY

Morality usually refers to that huge domain that treats how we should behave, what we hold as good or right and how best to achieve it. Although there is continuing debate over the scope of morality, we generally agree that, although it may have something to say about personal behavior, it concentrates on interpersonal relations. The domain of "oughts" is often contrasted with the world of nature that can be studied by science. If we believe in a sharp separation of the two worlds, we might confine the study of morality to religion or philosophy.

An entirely naturalistic study of morality—one that might leave everything in the domain of "what is," perhaps depending on evolution to select what is best for survival—seems unrealistic as well as unsatisfying. Given the enormous influence of cultural factors and the lengthy periods of time required to realize most biological-evolutionary changes, such an approach would negate the cognitive and affective advantages already achieved. It is necessary to take account of human choice and responsibility, and this opens another huge domain—one that cannot be limited to rational choice or some branch of game theory. We would make a mistake to assume that, once we have uncovered the roots of morality in evolutionary events, we now have the whole story on morality.[11]

Alasdair MacIntyre has argued that philosophers and others studying morality have made a mistake in moving away from rich descriptions and analysis of social life to technical analysis of moral statements, judgments, and universal principles.[12] I think he is right on this, but I will argue that an even greater mistake was made in ignoring female experience. Starting with maternal instinct, humans have developed groups and communities characterized by what I will call *natural caring*. It is that social state—one in which people respond naturally to one another's needs and feelings—that is prized by most human beings, and I will give considerable space to its description. *Natural caring* is not a conceptual contrivance. It is a state we see in everyday life—a practical, empathic mode of responding to one another. It is a social way of interacting with others, and we treasure it.

But it can be disrupted in innumerable ways, and then we need a more formal way of governing our responses. The purpose of *formal morality* from the perspective of care theory (what I will call *ethical caring* in chapter 3), then, is to sustain and expand the community of natural caring. Morality may be thought of as the set of attitudes, rules, and practices that aim to accomplish this purpose.

Consider how different this approach is from the technical work that occupies so many philosophers and psychologists today. The following example may help to make my point. Researchers studying morality present a problem to subjects to see what they will draw upon in making a moral decision. One such problem is known as the Trolley Dilemma. It is presented in two parts. In the first scenario, the subject is faced with a runaway trolley that will plow into five workers on the track. The subject may throw a switch that will redirect the trolley onto a siding where only one person will be killed. Would you (the subject) throw the switch?

In the second scenario, the same conditions exist, but no switch is involved. Instead, you are standing beside a very large man whose body—if he is pushed onto the track—will stop the trolley. Would you push him?

Investigators have found that most people would pull the switch in scenario 1, but they would not push the large stranger (thus killing him) to save five others in scenario 2. Ostensibly, this tells us something about the effects of proximity on our moral responses. We are not willing to kill even for a "good" purpose if we have to do it face-to-face.

Now my question is, Is the Trolley Dilemma really a moral problem? When I first read scenario 1, I thought, How could I throw a switch to change tracks? I don't know how to do this. Surely, one needs a special instrument? We can't have just anyone running about switching trains from one track to another. Then the influence of years of mathematics, logic, and philosophy kicked in, and I thought, Well, it's just an abstract problem—a game of sorts, and I should pretend I know how to do this. Maybe they are trying to count the number of Utilitarians versus the number of Kantians in the population.

But then I decided to think about the scenarios as many of the participants in Carol Gilligan's study did.[13] I began to consider the human beings involved as though this were a true story. Suppose the five workers are part of a chain gang of really despicable criminals. I would still prefer to save their lives, of course. But suppose the person on the sidetrack is our good mayor, or an innocent schoolboy, or my elderly mother. (Why in the world would either the mayor or my mother be prowling around a railroad siding? Never mind, play the game.) Surely I would not sacrifice any of the three to save five criminals.

What would I really do in either scenario? In scenario 1, I would yell, wave my arms, jump up and down to warn the men on the track, and in scenario 2, I would poke the stranger next to me so that he, too, in perhaps a louder voice, would join in the clamor of warning. I might clutch his arm in horror, but I would not push him.

I am not arguing that we learn nothing at all from these abstract dilemmas, nor am I arguing that nothing remotely like this ever faces us in the real world. Along with the Trolley Dilemma, the Lifeboat Dilemma has also been popular. In this story, people crammed into a lifeboat in violently stormy seas realize that their boat cannot sustain such a large passenger load for long, and they believe that, with no rescue on the way, they will have to make it to shore in the lifeboat. Unless they lighten the load, everyone will die. How should they lighten the load? In the nastiest versions of the dilemma, students are asked to evaluate the relative worth of the passengers' lives. Should we toss the elderly overboard? They don't have much longer to live anyway. But two very heavy men would do as much to lighten the load as three or four scrawny elders. In some versions, a saint—perhaps Mother Teresa—is included in the group. Would you value your own ordinary mother over the saint? To their great credit, I have known students who refused to play this game.

There was an account of a real-life situation in which crew members were faced with this dilemma, and they responded (logically) by chucking some passengers overboard. But, then, an unexpected wonder occurred. A vessel appeared and rescued the lifeboat's remaining passengers. Everyone could have been saved. This is a story of tragedy for both those sacrificed and those who sacrificed them.

It is sometimes said that subjects (usually women) who introduce the sort of extraneous material I described in my musings over the Trolley Dilemma just don't understand the nature of a moral/mathematical problem. One must accept the given and only the given and work one's way to a conclusion using well-defined principles. But people who argue this way should study mathematics more carefully. The history of mathematics is loaded with accounts of those who challenged the "given" in proposed theorems with counterexamples. Then mathematicians had to amend (sometimes several times) the "given" in order to preserve the conclusion.

My contention, however, is that moral problems should not be reduced to mathematical dilemmas. Life is too complicated for that. I am arguing that care ethics finds these abstract problems to be a distraction from the real, deeply felt problems of moral life. One might even argue that there is no truly *moral* solution to these dilemmas—only expedient, sometimes tragic solutions. A moral approach would concentrate on preventing such situations as nearly as possible. A guard or lookout would watch over workers on the tracks, an automatic device would halt a runaway trolley, parents would warn kids against playing on railroad tracks, ships would be equipped with sufficient lifeboats. In more realistic situations such as conflicts that sometimes culminate in violence, moral effort would be concentrated on preventing the conditions that inevitably lead to a loss of moral control. This is one of the aims of care ethics. The bottom line here is that a community of natural caring would fight the elements and accidents together. We should sensibly fear machinery run amok and raging storms, but we should not have to fear one another.

That something is logically amiss in these dilemmas can be confirmed by continually changing the given conditions. One who seems to be a Kantian in both the trolley and lifeboat situations may become a Utilitarian if faced with the choice of killing a handful of people to save an entire city from destruction. But the identification of Kantians and Utilitarians may be dead wrong. In the last situation, the decision maker (deciding to save the city) may be thinking of neither numbers nor absolute principles, but rather the destruction of circles of natural caring— helpless infants, terrified mothers, elderly parents, couples making love, pet dogs and cats, doctors saving lives, artifacts and mementos of generations, kettles boiling for tea, gardens full of vegetables, porch swings, and so on. Of course, the decision maker would do almost anything to preserve these circles of natural care but might not label the decision as *moral*, nor the opposite choice *immoral*. When things have reached this awful stage, the word *moral* loses its meaning. There is immorality here, certainly, but it lies in the creation of conditions that force people to make such decisions; the very possibility of morality is destroyed in such situations. Care ethics is aimed at preserving, constructing, and extending the conditions that eliminate or reduce the need for such decisions.

Quite a lot has been written on care ethics in the last twenty-five years, but there are still conceptual problems to address, and new problems have arisen as philosophers have explored the connections of caring to virtue ethics, justice, and the social/economic problems women face as caregivers.[14] All of these issues will be considered in later chapters.

First, we should consider some basic conceptual problems. Is *caring* a moral concept, or does it belong to some other social domain? Gilligan's *In a Different Voice* challenged philosophers and psychologists to revisit this question. Some critics responded in a manner directly derived from Immanuel Kant, insisting that ethics has to do with justice, impartiality, and moral judgments. Gilligan's work, they said, had to do with relationships and thus belongs to another social domain. This attitude has a long history. Kant claimed that women are naturally inclined toward the beautiful and, in social relations, to be kind and loving.[15] They had no need to think deeply, were really incapable of doing so, and should not engage in intellectually demanding activity. For Kant, a good act done out of love or inclination was admirable but had no moral standing. To be *moral*, an act had to be logically formulated and decided upon by reference to the appropriate principle. Lawrence Kohlberg, whose stage theory of moral development was challenged by Gilligan, followed Kant in making universal justice and the principles governing it the highest level of moral reasoning.[16] However, most people (including many philosophers) recognize that, although reason is certainly necessary in the moral domain, it does not have to be reasoning on stated principles, and emotions also have a place in discussion of morality. Gilligan's female subjects used reason, but they invoked it in assessing relationships and response. The point of view in this book is that all human interactions have a moral dimension, and both social and personal relationships are part of the moral domain of natural caring.

Positive reactions to Gilligan's identification of a different voice in the moral domain were astonishing. It was as though many women felt that their own voices were finally being heard. Her work was widely read and discussed not only by feminist scholars but by ordinary women all over the world.[17] Among feminist philosophers, one reaction was to point out that there were philosophers, such as David Hume, who had put great emphasis on the emotions and social virtues. It was time, they suggested, to revive this way of looking at moral philosophy.[18]

A troubling, if understandable, response came from other feminist thinkers who oppose the notion of *essentialism*, the idea that women and men were created with essentially different and unchanging natures. I'll say more about essentialism in chapter 2, where in the discussion of natural caring I'll suggest that we cannot reject the idea entirely. There are some characteristics that define human beings as a species, and most important, there may be characteristics whose absence would mean the extinction of human life. But for now it will be enough to see why so many feminists reject essentialism. Their reaction is understandable because the idea that women are essentially different from men has been used in the past to support the notion that women are inferior to men on the qualities most highly prized. No reasonable feminist would want to subscribe to such an idea. But one need not embrace essentialism to recognize that women and men differ on social attitudes and approaches to relationships. One can reject the Aristotelian essentialism shared by many religions and posit instead fairly stable differences induced by evolutionary experience. Such differences may be deeply embedded in our present "natures," but they are subject to further changes. Similarly, care theorists should reject the form of evolutionary biology that suggests there have been no changes in the human brain since the Stone Age. Biologists who talk this way sometimes refer to the difficulties of living in a modern world with a "Stone Age brain." But current genetic studies show that important changes have occurred and that some evolutionary changes take place over relatively short periods of time. We have to recognize that, somewhere along the line, culture began to play a role as large as that of biology in evolution, and the two are interactive.

The concern about essentialism is well taken, but it is wrong to attribute it, as some writers do, to Gilligan.[19] It has been around for centuries in philosophy, psychology, anthropology, fiction, and religion. Gilligan in fact made it clear that much of what she described as a different voice could be traced to socialization. My view is that the differences go beyond socialization; at least some can be traced to biological evolution. More needs to be said about these differences. Geneticists are daily finding important differences in genes at both the individual and group levels and in how genes trigger and are triggered by various bodily chemicals and by elements in the environment.

Commitment to an open mind on differences that seem well entrenched and that can be traced to evolution does not force us to accept an extreme form of reductionism. We can recognize that certain characteristics and tendencies have evolutionary roots without believing that everything can be explained this way. Indeed, part of what has developed through a combination of biological and cultural evolution is a human capacity to reflect upon and sometimes to change our own "nature."[20]

Rosalind Barnett and Caryl Rivers, writers who charge Gilligan with essentialism, also accuse her of baiting the "caring trap," encouraging women to return to the family life consonant with their essential nature.[21] If there were any truth in that accusation, we would expect to see a significant retreat among women from public to private life over the last twenty-five years. In fact, the opposite has happened. A newly realized sense of competence has impelled women to work effectively in the public world *and* to speak out more authoritatively on matters of caring. Philosophers have pointed out that, in developing the ethics of care, it is necessary to distinguish *caring*, which has application to all of moral life, from *caregiving*, which is an important form of labor that may or may not be accompanied by caring. This distinction will be important in everything that follows in this book.

We should also keep in mind the possibility that not all of moral life can be explained by one comprehensive theory. It is a temptation to take one's own favorite theory and stretch it to fit the whole moral domain. Recently, Michael Slote has made this claim for care ethics. He writes that his book "seeks to show that a care-ethical approach makes sense across the whole range of normative moral and political issues that philosophers have sought to deal with."[22] If he means that care ethics has something to contribute across all of the large domains concerned with morality, most care theorists would agree. I do not believe he is suggesting that the care-ethical approach can supplant all others; that would be a mistake. Such thinking is rather like that employed in utopian plans—one final blueprint for a nearly perfect society or way of life.[23] There may be no need to discard or replace other theories entirely. A more modest approach might claim biological primacy for care ethics and then show how other views—already articulated quite beautifully—are implicitly built on the foundation of caring. (This may indeed be what Slote has in mind, in which case we agree.) I argued this point in an earlier book and will extend the argument here.[24] There is no need, for example, to invent a whole new theory of justice if it can be shown that existing theories implicitly incorporate a notion of caring, but it may be necessary to inject specific elements of caring here and there as correctives.[25] In chapter 3, I will outline the beginnings of a care-driven approach to justice.

Care ethics is a needs-based approach, but that does not mean that it is uninterested in rights. Anyone living in a Western liberal democracy would be foolish to reject the notion of rights, but if we take an evolutionary perspective and stay close to reality, we have to agree with Jeremy Bentham that the doctrine of natural rights is "nonsense on stilts." Members of some organized societies grant one another certain rights, and they sometimes retract those same rights. The contribution of care ethics is to examine the needs and wants that underlie the granting of rights and to question an assumed right claimed by some when its exercise might deprive others of the satisfaction of needs.

The need to return to primary elements and to concrete cases will be addressed often in what follows. The temptation in theory building is to climb the ladders of abstraction, risking the loss of anything at all practical. To avoid this error, it is necessary to install *turning points*, points at which we return to origins in either theory or practice to see if what we are saying still makes sense and also to see where our own thinking should give way to good thinking already established.[26] Closely related to the need for turning points is a related need to stay close to reality. One objective, of course, is to construct and defend a workable normative ethic, but a prescriptive ethic should take adequate account of the way things are, including our inherited tendencies to care most about those closely related to us. In *Caring*, I gave an example of a woman who sided with her family at the barricades even though she believed their racist attitudes and acts were wrong.[27] Some critics interpreted the story as advocating family loyalty over what is morally right. But the story was *descriptive*, not *prescriptive*. One could, of course, argue for the priority of family loyalty along moral lines, but I was not doing that. Rather, I was arguing for (1) the necessity to recognize the way human beings—like other animals—care for their own, and (2) the necessity to think ahead and act with care so that these awful decisions do not become necessary—so that we do not find ourselves at the barricades.

Perhaps this is the place, again as a preview, to mention another special characteristic of care ethics. Like all normative ethics, it is concerned with the moral justification of acts and decisions, but its emphasis is not on justification, and it does not concentrate on moments of decision. It is concerned with moral life considered whole, with what precedes and follows particular moral decisions, with preventing conflicts of the sort alluded to above, and with what may be done to preserve caring relations when a decision unavoidably causes someone pain.

PRELIMINARY CONCEPTS AND DISTINCTIONS

It is necessary first to say quite a bit about caring and the several ways in which the term is used by care ethicists. *Caring*, in every approach, involves attention, empathic response, and a commitment to respond to legitimate needs. It is sometimes referred to as an attitude, but it is more than that; it is a set of dispositions to respond positively in interpersonal relations. Ethical caring involves a commitment. We will need to distinguish between caring-for and caring-about; between caring encounters, episodes, and long-term relations; and between natural caring and ethical caring. Much more will be said also about empathy, attention, response, biological response, cognitive apprehension, and the role of emotion.

The basic concept in care ethics as I develop it is the *caring relation*. Life itself starts in such a relation. The caring relation is an empirical reality, not a theoretical construct. Even before birth, there is a primitive caring relation, and the survival of the infant depends on its maintenance. Because the relation is taken as basic, the roles of both carer and cared-for must be described. In addition to the relations that are in a sense given—they are not chosen, we simply find ourselves in them—human beings construct relations. Recognizing that caring relations must be maintained and constructed, we can ask about the conditions that make it more likely that such relations will flourish.

Much of chapter 3 will be devoted to the practice of caregiving and its connection to caring. How do the demands of caregiving give rise to caring, and how is it that caregiving sometimes fails to call forth caring? In that chapter, we will also consider how *ethical* caring develops from *natural* caring, and we will trace natural caring to *instinctive* caring. Natural caring carries a large load of cultural influences, but it is still "natural" in the sense that it occurs out of love or inclination and does not require a moral effort; it is also natural inasmuch as it is observable in the empirical world. Although it may require considerable physical or emotional effort in the response stage, it does not have to be ethically summoned. Finally, ethical caring will be examined. In contrast to most moral theories, ethical caring will not be considered superior to natural caring. Indeed, I will argue that a primary purpose of ethical caring is to establish or restore natural caring relations.

In the discussion of caregiving, care as labor will be treated. When women persist in giving care, have they succumbed to the "caring trap"? Should young women be discouraged from entering the so-called caring professions? Is it feasible to work toward a rescaling of human values that would elevate those professions to a level consonant with their contribution to human welfare? This may be a wiser alternative than stamping our collective foot and insisting that since there are no real differences between the sexes, women should no longer do this work and instead should enter en masse into the better-paid professions formerly reserved for men.

Much has been written in the last few years about caring as a virtue. Some of this material compares care ethics with Confucianism, and some of it would make care ethics a branch of virtue ethics. There are similarities among these approaches to moral thought, but there are also important differences. Is caring a virtue? If it is, it must be described as a complex virtue, and it must be analyzed to find and describe its components. Moreover, there is a question of primacy. Virtue ethicists begin with caring as a virtue, but where does this virtue have its start? Care ethics, starting with the primal caring relation, construes that relation as the incubator or origin of virtue. When we have a clear picture of caring relations, we may describe a "caring person" as one who regularly establishes and maintains caring relations. Care ethics starts with a dyad, not with a lone, virtuous individual.

The discussion of virtue will lead to one of religion and its role in moral life. The question to ask and to probe is, What ethical need do women have for God or gods? And the answer will be that they have no such need, although some may have some other—nonethical—need for God. Women's acceptance of the wrathful, domineering, male God of the three great monotheisms is a product of their early and necessary acceptance of subordination. Recognizing this, women might do well to walk out on the religious institutions that continue to oppress them.

Have women failed to develop autonomy, or is there some other explanation for their continued subordination? In chapter 4, we will consider questions concerning autonomy. Is there such a thing as *autonomy*, or should we settle for intelligent heteronomy? What is gained and what is lost when we place autonomy at the center of social/political thinking?

Care ethics gives a central place to the role of emotions in moral life. The emotions, as Hume demonstrated, are basic in human action. In chapter 6, I will show that a close examination of the female evolutionary path supports Hume's argument. But we do not act from blind emotion, and—using reason well— we cultivate the moral sentiments. Careful, objective observers can see that males are almost certainly more emotional than females—leaping for joy at victories in sports, shouting and waving fists in political protests, attacking one another in road rage, celebrating the brutalities of combat. Then, too, dominant emotions differ in males and females. Whereas men are likely to give way to anger and physical arousal, women are more inclined to cry in sympathy or frustration, and an evolutionary account helps to explain the differences in emotional display. The inner, reasoned voice of females counsels against anger and bravado; the male inner voice often uses reason to justify anger and violence.

In chapter 7, "Needs, Wants, and Interests," I will extend arguments I made in *Starting at Home* and connect them to the evolutionary line of thought. The social environment, universally, has demanded that women pay attention to *expressed* needs, whereas religion and schooling have built more closely on *assumed* needs.[28] This discussion offers an opportunity to more fully develop the contrast between virtue ethics and care ethics and to analyze the important differences between virtue-caring and relational-caring.

Is a female approach to moral life more likely to promote peace than the moral philosophies so far developed?[29] That will be the topic of chapter 8. The peace ethic growing out of care ethics is not an idealistic, pacifist theory. Females know that they will fight for the lives of their children and that—with great sorrow—they will side with their own in times of conflict, even when their own are demonstrably wrong. This is a descriptive fact, compatible with evolution theory, and a defensible peace ethic must take it into account. All the attempts to regulate and justify war and its conduct are hopeless. Our efforts must concentrate on eliminating or reducing the conditions that contribute to war and violence.

In the final chapter, I explore whether ethical convergence is possible. There are many promising and beautiful ideas in moral philosophy as it has developed in the male line, and they could be enriched by those emerging in the female line. I have already expressed some agreement with Alasdair MacIntyre that moral philosophy went wrong in separating itself from social life more broadly considered. But it went wrong even earlier with Plato, Aristotle, and the religious theorists who embraced monotheism; it went badly wrong by ignoring or even deploring female experience. The possible benefits of convergence will be discussed. However, forces working against convergence are enormously powerful, and a clear-eyed assessment does not encourage optimism. In any case, readers should understand that I am exploring one significant source of morality—maternal instinct and the natural caring that develops from it. I do not claim that it is the only source of morality.

TWO

The Caring Relation

Many feminist philosophers view human beings as relational selves—selves constituted by the relations in which they are embedded.[1] For care theorists, the *caring relation* is morally basic, and it is the purpose of this chapter to describe its development from the original caring relation established by maternal instinct to natural caring. The last section of the chapter will explore how we might best prepare people to establish and maintain caring relations.

INSTINCTIVE AND NATURAL CARING

In exploring a path to morality rooted in maternal caring, we must look at instinctive caring, natural caring, and ethical caring, but these should not be considered stages in moral development. Certainly natural caring has incorporated instinctive caring and, because it seems to have evolved from instinct, it represents a next step. But ethical caring, which again builds on natural caring, is not necessarily better than natural caring. It becomes morally necessary when natural caring fails in its usual settings or when the setting has become too large for natural caring to function. In the latter case, it seems reasonable to many that we should switch from caring to justice. But a concept of justice may be guided by the fundamental ideas of caring. The best approach for policymakers might then be to choose a theory of justice that aims to establish or restore conditions in which natural caring might flourish rather than to invoke an entirely different moral approach or to attempt caring directly. Throughout the discussion, the use of *caring* will have to be clarified repeatedly.

Every human being is born into, takes shape in, relation. For the embryo and fetus, life exists only in relation. Physical separation at birth initiates a caring relation once maintained almost entirely by instinct. Since the emergence of recognized human life, the whole process has been overlaid with practices developed in cultural evolution and, as part of cognitive evolution, with the conscious attention of mothers capable of thoughtful planning and reflection.[2]

The dyadic connection consisting of mother and child was originally one of instinctive caring. We might think of this caring relation between mother and child as "naturally normative," in the language of Philippa Foot.[3] Today, a woman who does not wish to bear children is not—and should not be— thought to be somehow defective, but a mother who does not care for her child might be considered as having a defect in the sense that she lacks an instinct or trait considered essential for the survival of the human species. By today's standards, such a woman is nevertheless—and I think rightly—regarded as fully human, because she possesses other qualities characteristic of the human species. Neither would we regard her as a "defective female," although we acknowledge this one defect. Cases of this sort underscore why so many feminists reject essentialism. Too much has been loaded onto one trait. However, if a significant number of females were to lack the characteristic of maternal caring, not only would that lack be labeled a defect in the individuals; the whole group would be seen as violating a natural norm—a defect that would predict the extinction of the species.

In this brief discussion of maternal instinct, we see that complete rejection of essentialism may not be possible. On the one hand, maternal instinct is not an essential characteristic of human females, one that separates fully human females from "unnatural" females; on the other hand, it is an essential characteristic of human females as a class in that it is clearly essential to the survival of their species. Humans are complex creatures, and we have learned to avoid labeling any infant born to a human mother "defective" because of some lack or anomaly. A child born with four arms, for example, is still human, and the anomaly (deviation from the natural norm) may be corrected. A severe mental deviation (what was once called idiocy or imbecility) is not correctable, but we point to other qualities that such a child shares with fellow humans. Again, if this deviation were widespread in a group, we would have to label it a severe violation of natural normativity because the group lacks a characteristic (normal human intelligence) essential to the continuation of human life.

The characteristic attitude of caring (we'll say much more about what this is) is manifested in love for the child—an almost fierce love inherited from its beginnings in maternal instinct. This first caring relation is important not only because human reproduction depends on it but also because it represents a prototype of caring. Except in rare cases when a mother emotionally rejects her infant, no moral imperative is involved in the mother's decision to care for her child. The caring is *natural* in this sense—that it is practiced out of love or inclination. Recall that, in Kant's ethical theory, acts done from love or inclination have no moral worth. To have moral worth, from Kant's perspective, an act must be chosen in obedience to an ethical principle. Mothers do not refer to a moral principle in deciding to feed their babies. They *want* to do so. Does their love or instinctive inclination have no moral worth?

I agree with Kant that there is a difference between natural caring and ethical caring, but I will argue that both have moral worth. We should distinguish between informal or everyday morality and formal morality. When I say that natural caring represents a moral approach to life, I am referring to an informal morality, a way of interacting with others that does not require explicit attention to moral criteria such as duty, principle, God's will, or the exercise of virtue. In a climate of natural caring, our attention is on identifying and satisfying needs, maintaining or establishing caring relations, keeping open the channels of sympathy, and promoting this way of life among the young. In care theory, we are concerned with the full scope of moral life, not only with ethical decisions and their justification. Natural caring is, in some ways, superior to ethical caring. Consider how we would feel if a friend were to visit us while we were ill and tell us frankly that he was doing so because it is his duty. We might well feel hurt. We might even wish he would simply go home. In almost all close relationships, many of the most important acts and attitudes are governed by inclination, not duty. Often, as we shall see, the purpose of ethical caring is to establish or restore natural caring. If natural caring never failed, if it could be extended without limit to all others, we would have no need for ethical caring.

The first caring relation is our original condition. The infant does not choose to be born and does not choose the family or community into which she or he is born. The child will eventually make choices, and this capacity to choose has led liberal philosophers (and some existentialists) to exaggerate the human capacity to choose. In the usual course of life, people do not choose their sex, race, or ethnicity. They do not choose their stature, physical strength, or susceptibility to disease. Indeed, few choose their religion, and many do not even choose their occupational way of life. It is extremely difficult to make choices in opposition to one's immediate culture. The liberal emphasis on the individual and individual choice has begun to give way, thanks to evolutionary studies, to an acknowledgment of the group or society as a unit that functions in much the way a "population" does in nonhuman species.[4] The social group is primary, and individuals are both developed and limited by it.

Evolutionary studies are having effects across the disciplines. Not only is emphasis shifting from the individual to social units in biology, anthropology, and psychology, but history is moving well beyond the "great man" as a center of interest.[5] Jean Piaget noted forty years ago that such thinking should change: "The great man who at any time seems to be launching some new line of thought is simply the point of intersection or synthesis of ideas which have been elaborated by a continuous process of cooperation, and, even if he is opposed to current opinions, he represents a response to underlying needs which arise outside himself. This is why the social environment is able to do so effectively for the intelligence what genetic recombinations of the population did for evolutionary variation or the transindividual cycle of the instincts."[6]

Natural caring is "natural" in the sense that it is exercised with no need for reference to moral principles or direct reasoning from such principles. Perhaps the easiest way to get an initial grasp on the difference between natural caring as a way of moral life and the body of thought traditionally associated with moral philosophy or ethics is to reflect on the "different voice" identified by Carol Gilligan. Instead of drawing on principles and rules, the female participants in Gilligan's study concentrated on relationships and response.[7] I am trying here to describe this alternative mode of moral life more fully.

Natural caring is not a stage—either historical or developmental—on the road to a "higher" morality. It has always been present and active in human life but largely unarticulated and overwhelmed by male-dominated theories of morality. These theories have concentrated on fundamental principles, often entirely separated from real (biological/social) life, ostensibly "anchored" in a mysterious entity or entities beyond that real life. Or they have depended on customs and rules developed in various occupations, practices, and traditions—all defined in terms of male experience. The development of male moral theories has tended toward the abstract with little attention to the facts of biological life. At the extreme, moral theory has become a sophisticated linguistic/mathematical game entirely foreign to the everyday social life of human beings. Indeed, because natural caring is visible in the actual lived world, it might be (has been) identified with "isness." In contrast, traditional philosophers have described ethical conduct (formal morality) in terms of "oughtness."

It has been that way from the beginning of male moral thought. Consider the culture in which classical Greek thought emerged. It was a culture constantly at war, and many of its virtues were defined as qualities of the warrior. It supported slavery, and its moral rules, customs, and virtues were tightly connected to social/political status. It denigrated women and women's contributions to family life, and it even made an attempt (through Plato) to show that family life was unnecessary. Where something practical was acknowledged and described (as in Aristotle), it arose in and was elaborated through male experience.

This is not to say that centuries of male thought on moral life have been useless. There are beautiful, inspiring ideas that continue to excite thought. My complaint is that they have had relatively little effect on actual human conduct. Instead of effecting an end to war, they have constructed elaborate schemes for the "moral" conduct of war. They have condemned slavery and contributed to its formal elimination in much of the world, but human trafficking continues, and millions still exist in slavelike conditions. Male philosophies still denigrate women by ignoring the contributions women have made *as women*. Forced by their own theories of justice to accept women as equals, they implicitly reemphasize the dominant status of men by inviting women to become more like men.

In his interesting exploration of the evolution of morality, Marc Hauser argues—as I do in this book—that we need to pay more attention to the natural origins of morality and to the moral feelings he calls "intuitions," but he then decides to move away from empirical reality in human life: "To understand our moral psychology, I will *not* explore all the ways in which we use it in our daily interactions with others. In the same way that linguists in the Chomskyan tradition sidestep issues of language use, focusing instead on the unconscious knowledge that gives each of us the competence to express and judge a limitless number of sentences, I adopt a similarly narrow focus with respect to morality. The result is a richly detailed explanation of how an unconscious and universal moral grammar underlies our judgments of right and wrong."[8] The result, from the perspective of care theory, is unnaturally narrow. Hauser does not mention female experience or maternal instinct. When he writes about the nurture of infants, he speaks of "parenting"—as though males and females play similar and equal roles. There is no reference at all to work in feminist psychology or philosophy. What is explained in absorbing detail is a notion of reciprocity that concentrates on the expectation of compensation or giving back. When a person does something for another, the expectation is that the other—perhaps at a later time—will "reciprocate"—do something helpful in return.

This way of looking at the origins of moral life is part of a parallel (and enormously powerful) philosophical tradition that puts the individual at the center of moral thought and action, building a system of ethics on the foundation of self-interest. The resulting description of moral life seems accurate when we look at the way things operate in the male-dominated world.

But *think*. Reciprocity in female experience often involves no expectation of compensatory action. A mother hopes for *response* from her infant, but there is nothing in her conduct that corresponds to contractual reciprocity—"you scratch my back, I'll scratch yours." Right from the start, we are talking about relation and, in the mother-child relation, the child comes first in the mother's thinking. In many of the interactions characteristic of female experience, the so-called conditions of reciprocal altruism do not apply. Mothers have not negotiated small costs for their giving, looking for great benefits in return, nor have they made giving contingent upon receiving.

The Golden Rule, heralded as a universal judgment, is admirable, but it too might be thought of as a male universal because, again, it puts the individual moral agent at the center: Do unto others as you would have done unto you. If we look at female life and experience, we might enunciate a somewhat different rule: Do unto others as they would have done unto them. This way of response requires many of the attitudes and skills described in an account of natural caring.

In later chapters, I'll discuss how care theory can provide guidance for social policy. Of course, in liberal democracies, we want to educate people so that they can make intelligent choices, but—even more important—we should work toward a society in which the original condition of every child is one in which natural caring can flourish.

Natural caring starts in the settings influenced by maternal instinct. The mother's care is lavished upon children, children are taught to care for one another, and the biological tendency to care for blood relations is augmented by social customs. However, natural caring is not limited to families. Wherever people come together out of common interests, we are likely to find natural caring—forms of address and response not motivated by rule or principle. Once again, I must emphasize that we may certainly detect and describe behavior that seems to be guided by rule or custom. It is descriptively correct to say that people in such communities behave in ways that can be predicted by descriptive principles. It would be a mistake, however, to say that these people refer to these rules before acting. Circles of common interest, guided by natural caring, can provide a foundation for a care-driven approach to justice, and I will say more about this in a later chapter.

Just as the entities we call *individuals* are initially defined by relations in the home, they continue to be shaped in various groups of common interest. In settings characterized by natural caring, people do not consciously refer to rules or principles to guide their behavior. Loved, or at least accepted, people go about their daily lives without recourse to explicit moral thought. They use reason, of course, primarily in means-ends analysis. How can I accomplish this end I seek? How can I reshape the mistaken goals of the cared-for?

A setting characterized by natural caring is widely (perhaps universally) regarded as *good*. It is thought *good* to be safe, to have one's needs met, to expect and receive response to one's overtures. The appreciation of natural caring is rooted deeply in our evolutionary past, but it is still an important feature of ordinary contemporary life. To return home after a hectic day's public work, kick off one's shoes, settle back with a sigh, cuddle a loved child (or pet), close one's eyes in recognition that one is safely situated in a circle of caring is to realize a basic human good.

Evaluating the conditions of natural caring as *good* gives rise to what we now think of as moral thought. If something goes wrong, how can I restore the conditions necessary to sustain relations of natural caring? When we ask the moral question, What ought I to do? we are usually asking what we would do if we were at our best as caring selves and if this other were one we naturally care for. In most such cases, there is no need to consult abstract rules and principles.

Although natural caring grows out of instinctive caring, it is clearly not merely instinctive. All mammalian females are capable of some rudimentary thought, and they exercise it in seeking an appropriate place to give birth, in choosing relatively safe times to search for food, and in deciding who may or may not approach their young. But human females are capable of fully human cognition. Moral reasoning is not involved in the decision to care, but reasoning is employed in deciding how best to satisfy the inclination to care.

Female humans, like virtually all mammalian females, have had and continue to have major responsibility for mothering, but human female thinking is not confined to the tasks of mothering. I argue in this book that it is reasonable to suppose that female and male minds have evolved somewhat differently, but I do not suppose that one is generally superior to the other. However, one may be better adapted than the other to certain tasks. This is a risky claim because, as many feminists have warned, admission of difference in the past has almost always resulted in a declaration of superiority favoring the male.[9] That arrogant conclusion is to be deplored, but it should not lead us to deny that differences exist. Rather, we should ask how best to acknowledge and use the differences to benefit everyone. In some cases, it might be beneficial to reduce the differences. *If* a difference exists between the sexes in mathematical ability, for example, should we work to reduce the difference? Why? *If* a difference exists in the capacity for empathy, should we work to reduce that difference?

Some care theorists concentrate on caring as a virtue and deemphasize care as a relation.[10] From an evolutionary perspective, however, the caring relation is primary. Indeed, it may be considered the incubator of virtue—phylogenetically, of the virtues associated with maternal care, ontogenetically, of the virtues developed in interaction between mother and child. This is not to deny other possible sources of virtue such as cooperative hunting and pathfinding among males. But the original caring relation is almost certainly a major source of empathy and the virtues associated with it.

The original caring relation is not chosen; it happens. Other caring relations may be constructed, and that possibility will occupy much of what follows. What can we say, for example, of the earliest male-female bondings? We know little about these. Why did females enter such relations? The relation itself (a mere association over some time) might not initially have been one of caring. As suggested earlier, the female's motive might have been concern for the survival of her young. However, the relation—dyadic interaction—might well give rise to caring in either partner. Males might become genuinely concerned with the welfare of their female partners and, secondarily but less intensely, with the survival of their offspring. Females, learning of necessity to read their male partners, might begin to care for them. From the merely instrumental, caring relations might develop. It seems more likely that females, with the growing capacity for empathy engendered by maternal experience, would nurture the transition from instrumental relation to caring relation. Or it just might happen that the female begins to feel something for her mate, and so what started as instrumental caregiving becomes natural caring. Neither instrumental caregiving nor natural caring requires reference to a principle.

Natural caring is "natural" in that it exists prior to formal moral thought; it is there, in the empirical world. It is found in families and in other face-to-face circles of interaction. As an organic system, it is characterized by mutual protectiveness, cooperation, response to one another's needs, and support for one another's projects. I am not talking about a moral utopia here. Although natural caring is usually found in family and small group situations, there are such groups (usually on their way to extinction) in which natural caring is absent, and in most groups natural caring sometimes fails. We might call groups that regularly fail to exhibit natural caring defective; they are lacking in essential human qualities. When natural caring fails episodically in groups that usually live by it, it is sorely missed. Everyone is thrown off balance, perhaps even deeply unhappy. On these occasions, ethical caring is required to restore natural caring.

Relational care theories recognize that the "individual" extolled by liberal philosophies is a myth. Life begins and develops in relation. The quality and quantity of interactions steer cultural evolution. Without denying other sources of morality, it seems reasonable to suggest that one powerful source is the maternal instinct that, with cognitive development, grew into natural caring and the virtues often associated with it. Right from the start, we are relational beings. As the relation is basic to biological life, the caring relation is basic to moral life. But how should this caring relation be described? What are its elements and what virtues are associated with it?

CARER AND CARED-FOR

We may identify two parties in a caring relation: a carer and a cared-for. Each contributes distinctively to the relation. Note that this marks an important difference between care theory and traditional moral philosophy, which usually concentrates on the individual moral agent and the agent's connection to principle or to the agent's own virtue. The original caring relation is what Martin Buber described as an *unequal* relation; that is, it is one in which the parties cannot exchange positions.[11] In this relation, the mother is necessarily the one caring (or the carer) and the infant the one cared-for. The same sort of inequality characterizes formally defined relationships such as physician-patient, teacher-student, and lawyer-client.

In most adult relationships, however, we expect mutuality; the parties exchange places as the situation within which the relation exists changes. Indeed, we often judge the health of long-term relationships by the number and appropriateness of these exchanges. It should not be expected that, in a female-male relationship, the female will always serve as carer and the male as cared-for. Some feminists have expressed worry that acceptance of care ethics might contribute to the long-standing exploitation of women—that women will fall into or fail to escape from the "caring trap."

This worry has some legitimacy, but the legitimacy rests on two mistakes: first, that "carer" applies permanently to a person by virtue of her gender, and second, that *caring* as it is used in care theory is identical with caregiving. If we eliminate these two misunderstandings, there should be no fear that care theory will set a trap for women. Notice that I do not deny the reality of a caring trap, and we'll have to discuss how it was set and how it continues to be baited.[12] But I do deny that care theory, properly understood, contributes to the maintenance of the trap.

Let's use the original caring relation as a setting in which to analyze the contributions of carer and cared-for. The carer must first of all be attentive to the expressed needs of the cared-for. These needs are not always expressed verbally. Obviously, an infant does not express itself in words, and there are many other situations in which needs are expressed with body language. The carer's task is to discern correctly what is being expressed. This is another topic on which much more will be said in chapter 7. For now, it should be enough to note that a carer must exercise receptive attention—to listen and watch.

Analysis of receptive attention has been relatively scarce in philosophy. Simone Weil discussed this form of attention and its place in moral life. The basic attitude of one exercising receptive attention is captured in the question asked (explicitly or implicitly) by carers: What are you going through? Weil writes: "This way of looking is first of all attentive. The soul empties itself of all its own contents in order to receive the being it is looking at, just as he is, in all his truth."[13]

In *Caring*, I used the word *engrossment* to name this form of attention, but the word has been too often misinterpreted.[14] I did not intend to suggest that an individual (carer) should be engrossed in another individual (cared-for) as a lover might be engrossed in the beloved. What I meant to suggest is that the carer is engrossed in (or receptively attentive to) the needs expressed in an encounter. In the case of maternal caring, the attention is continual; engrossment may be the right word. But in the momentary encounters we regularly experience—for example, being asked by a stranger for directions—we are engrossed or completely attentive just long enough to hear the request; we respond, and the moment passes. At these moments, as Buber put it, the other "fills the firmament."[15] But the stranger fills the firmament briefly; the infant fills the firmament often and not only when immediately present.

When one engages in receptive attention, the result is often motivational displacement; that is, the motive energy of the carer flows toward the needs of the cared-for. Temporarily, the carer's own projects are set aside. A teacher may feel her own fingers twitching as though they were holding a pencil as she gets ready to respond to a student's struggle with a math problem, and many parents have experienced a similar bodily reaction when watching a child's first wobbly performance on a bicycle. But motivational displacement is not always revealed in bodily reactions. Its fundamental character is a shift in the direction of motive energy.

Then, of course, the one-caring must do something. Decisions on how to respond—what to do—depend not only on the expressed need but also on the competence of the carer and the resources she has at her disposal.

An incipient caring encounter may be disrupted or aborted at any stage: the carer may fail to engage in receptive attention, the expressed need may not trigger motivational displacement, or the carer's response may not be recognized as caring by the cared-for. To qualify as a caring encounter, the cared-for must contribute to the relation; the cared-for must show in some way that the caring has been received. This response completes the caring relation. Without it, we may acknowledge a carer's effort to care, but the encounter, episode, or relation cannot be said to be one of caring. However, acknowledgment of the carer's efforts need not take the form of gratitude. An infant's smile, eye contact, and extended arms all represent acknowledgment. Similarly, a patient's release from pain or a student's renewed energy in pursuing a learning objective conveys to the carer that the caring has been received. Further, the response of the cared-for supplies more information for the carer to use in the next encounter. The response is vital to the caring relation. Requiring a response of recognition from the cared-for makes care ethics distinct from virtue ethics. We need not interpret the infant's smile, the patient's sigh of relief, or the student's perseverance as virtuous, but we recognize that these responses contribute to the caring relation and, more generally, to moral life. In care ethics, *caring* more often points to the quality of relation than to a virtue in the one caring.

I should clarify use of the words *encounter, episode,* and *relation.* All three point to dyadic interactions. An *encounter* might be thought of as a minimal relation (A, B) in which A acts as carer and B as cared-for. B expresses some need—for example, for directions or help in lifting a heavy package—and A responds by meeting the need or, at least, by explaining regretfully why she cannot do so. A may, for example, reply, "I'm a stranger here myself," or "I'm just recovering from a back injury," and may help by trying to enlist a third person's aid. A has been attentive, felt the desire to help, and responded with an expression of concern; B has recognized A's effort as caring. The encounter is rightly called a (minimal) caring relation.

By *episode,* I mean an encounter or set of encounters within a longer-term relation. A long-term relation such as mother-child is a caring relation if most of the episodes within it are caring and the growing child acknowledges the continuing relation as caring. Such relations are also characterized by continuing interest or concern between encounters. Mothers do not forget about their children when they are not interacting directly with them, nor do good teachers abandon concern for their students between classes. Carers in long-term caring relations maintain a disposition to care; they are prepared to care. In the last section of this chapter, we'll examine this idea more closely. How is it that some people approach the world "prepared to care," and how do we cultivate this attitude?

Discussion of being prepared to care necessitates examination of another important distinction—that between *caring for* and *caring about*. *Caring for* requires direct interaction with the cared-for, the criteria of caring already laid out, and the expectation of response from the cared-for showing the caring has been received. In contrast, *caring about* is characterized by concern, perhaps about people at a distance or about groups of people at some risk. Although we are not always in a position to *care for* people, when we sincerely *care about* them, we may contribute to a charity, vote in a way thought to improve their condition, or moderate our own way of life in an effort to conserve resources.

Caring about may provide one foundation for a theory of justice. (Another, of course, is self-interest.) We cannot care for everyone; that is physically impossible. When there is no provision for direct encounter and reception of the all-important response of the cared-for, we try to employ some form of justice in our policies. Justice does not become irrelevant when we embrace an ethic of care. However, care theory counsels that whatever is done in the name of justice be guided in part by the spirit of caring. Instead of imposing rules and practices derived from our own conception of justice, we ask how we might help to establish conditions in which caring relations can flourish in the society or culture that will receive our care/justice. Thus the caring relation and caring-for remain basic in both theory and practice.

Often in caring about, as in caring for, we are moved by empathy, but the intensity of feeling is not usually so strong. We contribute something to overseas needs, but we continue to give most of our attention to those close to us. Occasionally, when feeling is very strong, a person may drop everything else and seek a way to establish direct encounters with those cared about. Stories about people who do this excite considerable public interest, but we recognize that few people are in a position to do this. However, we note that *caring about* may inspire *caring for*.

EMPATHY

In a previous section, I considered the use of *empathy* to describe what motivates someone to care. We must be careful, however, not to use the early, projective definition of empathy. Contrary to its first uses in aesthetics, *empathy* is not now identified totally with the intellectual. Indeed, as it is used by Martin Hoffman and Michael Slote, it is closer to the traditional definition of sympathy. As they use it, it has both cognitive and emotional dimensions. But what exactly are the processes involved in empathy?

Attention is certainly involved in empathy. As noted earlier, Simone Weil put great emphasis on attention. Her comment on the soul emptying "itself of all its own contents in order to receive the being it is looking at" comes close to capturing the receptive attention required in caring for another. But we can never really empty ourselves of our interests and values. If attending, receiving what is there generates sympathy and moves us to motivational displacement, we do put our own interests and projects aside temporarily, but they are still present—at the side, so to speak—and may be invoked in deciding how we should respond.

Lovely as Weil's description of attention is, we may not want to follow her too closely. First, she suggests that the faculty for attention may be developed through school studies such as geometry.[16] But the sort of attention applied to geometry may not be the sort required in caring; many mathematicians have poorly developed skills in reading people. Studies of mathematical intuition do sometimes portray mathematical attention as receptive, and some mathematicians have even been described as being "seized by mathematics." In such cases, mathematics "fills the firmament" and "the soul empties itself" of all non-mathematical content. So perhaps the difference lies entirely in the object of attention. If so, and if our objective is to develop the capacity for attention to people, we face the problem of transfer—one familiar to psychologists who study learning. Transfer of learning typically takes place over a narrow range of similar topics or problems. Another unhappy possibility is that receptive attention regularly exercised in one domain leaves little energy for its exercise in another. I'll say more about this in a discussion of caregiving. In any case, there is no evidence that studying geometry increases one's capacity for attention to persons.

Second, Weil invokes an intermediary, God, in the process. Students supposedly study geometry and thereby increase their capacity for attention, attention is then directed toward loving God, and that love is then passed on in attention to fellow human beings. The attention described in care theory does not require an intermediary; it goes directly from carer to cared-for without the mediation of God or principle. In care theory, anything that comes between carer and cared-for must be examined closely. Often, it acts as a distraction.

Weil's perspective is far different from care theory when she discusses cries of pain and protests against injustice. For her, the only important reaction to evil is impersonal. The protesting cry of pain must somehow be transcendent, not personal. She writes: "There are also many cries of personal protests, but they are unimportant; you may provoke as many of them as you wish without violating anything sacred. So far from its being his person, what is sacred in a human being is the impersonal in him."[17]

For care theorists, however, it is the living other in all his or her personal, embodied distinctiveness who addresses us. The other is *not* "exactly like us," as Weil would have it. Indeed, there would be little need for the attention Weil describes so beautifully if we were all alike. There would be no need for the basic question, What are you going through?

In contrast to Weil's emphasis on the impersonal and transcendent, Iris Murdoch, in her analysis of attention, says, "Love is knowledge of the individual."[18] But this knowledge of the individual is sought both through cognitive apprehension and through the viewer's habit of being prepared to care. Cognitive apprehension is the "reading" process already discussed. In the primal relation, it is motivated by instinctive love. In early woman-man pairs, its accuracy was probably improved by fear—apprehension as fear or dread. With evolution in the race and development in individuals, the accuracy of apprehension grows and is supplemented by reflection, which encourages the one-caring to evaluate her perception and to look again.[19]

Murdoch uses the case of M, a mother-in-law who sees her son's wife, D, as "unpolished and lacking in dignity and refinement."[20] Despite her initial reading, M behaves beautifully toward D, but she feels uncomfortable because her behavior is not genuine—that is, it does not match what she really feels. M thinks about this relationship. She reflects on her own attitudes and criticizes herself for being old-fashioned, conventional, perhaps even narrow-minded and snobbish, and she decides to "look again" at D. Murdoch comments: "What M . . . is attempting to do is not just to see D accurately but to see her justly or lovingly."[21] M's behavior does not change (Murdoch supposes), but something surely has changed.

From the perspective of care theory, M initially drew on ethical caring to guide her behavior. She behaved toward D as she would have if she cared naturally for her. But, as we'll see, there is always something risky about this. We can hypothesize, with Murdoch, that there was no discernible difference between M's earlier and later behavior, but in real life, all sorts of differences might be detectable—a veiled glance, raised eyebrow, or forced smile detected by D when M was unaware of coming under D's gaze. D might in fact have mistrusted M's caring, and then despite appearances, there would be no caring relation. M evidently anticipated something like this in the situation and drew upon past experience of caring to reflect and then to work at establishing a caring relation.

At what point did M feel empathy for D? We can assume that she saw D's behavior accurately, but she realized that different valuations might be put on the behavior, and she decided to look again through a different, more caring lens. Why? Perhaps something the girl did or said triggered an initial sympathy. Perhaps she made the decision out of love for her son. Or perhaps she referred to her own ideal of caring. She used ethical caring to work toward the establishment of natural caring.

Almost certainly, M felt considerable sympathy for D, and her sympathy was the motive force for reevaluating her own judgments. Here my original reluctance to depend on empathy returns. Empathy without sympathy may aggravate an already unfavorable judgment. Sometimes the more we know and understand another, the less we like the other, who may be quite thoroughly rotten, or the fault—a fault of personality—may lie in ourselves. Michael Stocker remarks:

> Our overall conclusion, then, is that it is quite often false that knowledge of others will produce sympathy and liking. In many people, for many different reasons, knowledge produces dislike and hatred, and more knowledge produces even more dislike and hatred.
>
> It thus looks as if empathy will lead to sympathy only for those who are already sympathetic.[22]

But there is another way of looking at the picture. Even when empathy leads to dislike or disgust, sympathy may be triggered by an expression of fear or pain in the one for whom we have no liking. Empathy does not ensure sympathy, but sympathy is not entirely dependent on empathy.

We can, of course, eliminate this possibility by fiat—by insisting that sympathy be built into the definition of empathy. For the most part, I will follow this path, but whenever the possibility of conflict arises, I will revert to the specific use of *sympathy*. In chapter 6, I will say much more about sympathetic attention as a virtue.

Empathy, then, is a constellation of processes. It involves attention; cognitive apprehension, or reading (the results of which may or may not be reevaluated); a strong possibility of sympathy; and connection to one's own sympathetic structures. What gets these processes started? There seems to be some circularity here. In some cases, empathy begets empathy; that is, someone who regularly experiences the processes above is already "prepared to care" and approaches encounters with an empathic disposition. Sometimes love motivates the process, but we know that love may interfere with empathic accuracy; lovers often see the beloved in a rosy light that dims over time. Sometimes some feature of the situation or of the cared-for triggers memories in the carer, and the empathic process begins. And sometimes immediate sympathy gets the whole thing started.

In the case of Murdoch's M, there is no reference to principle, and there is no attempt at impartiality. Lawrence Blum points out that traditional moral theories too often dismiss the moral approach described by Murdoch.[23] In the language I am using here, they miss the whole realm of natural caring. In doing so, they also miss one of the most powerful motives for being moral—to restore the conditions of natural caring that we regard as good.

In the original caring relation, maternal love instigates the processes. At the phylogenetic level, selection would favor females with more highly developed empathic capacity, and at the individual level, motherhood or maternal thinking might promote further empathic development.[24]

With the understanding that *empathy* includes (at least) the processes described here, I will use it as a shorthand in discussing preparation to care. We can equate the development of empathy with growth in the capacity to care—that is, with the development of people who are prepared to care. In everything that follows, however, the emotional element of empathy will be the driving force. Cognitive apprehension is essential, but so is the motivation to look again or reevaluate—to attempt to see lovingly. On the other side, there is a cognitive contribution connecting sympathy and motivational displacement. Carers reflect on their sympathetic reactions and either permit or reject the move to motivational displacement.

Let's return briefly to the idea that grounds development of an ethic of care. Are women more empathic than men? Michael Slote even claims that women are more moral than men, but I will hold off on this. Evidence currently available suggests that women are more empathic than men, and this should not surprise us, given the thousands of years of female caregiving experience with which evolution has had to work. But remember that their increased capacity for empathy has come at a cost—acceptance of subordination and sometimes enthusiastic endorsement of their own subservience.

Another note of caution is needed here. I have raised a caution about essentialism. I do not believe that women were created with an eternal, unchangeable nature. An evolutionary view has to accept change. Selection usually works to increase survival, but there is now plenty of evidence that some mutations bring ills along with the improvements for which they seem to aim, and I am not claiming that women's superior capacity for empathy makes them morally superior. Other factors are involved.

In the next few decades, geneticists are almost sure to find more evidence of genetic and chemical/hereditary influences on behavior.[25] This research will often be heatedly challenged, because it seems to call into question much of what we would like to believe about ourselves and our social/political ideals. I will continue to argue that we can face the new information—much of it staring us in the face daily—and find new and more realistic ways to promote a more just and caring world.

PREPARING TO CARE

The mother-child relation, as the original condition, is the primary example of natural caring, but unlike other relations of natural caring, it still has firm roots in instinct. Most of our interpersonal relations require the construction of caring relations. There is some evidence that there may be an element of instinct in the way we behave altruistically toward blood relations, but caring even among those closely related requires maintenance. Sometimes things go wrong in the domain of natural caring, and we have to draw upon ethical caring in order to restore natural caring relations. Ethical caring will be the focus of our attention in chapter 3.

In this section, I want to consider how natural caring is extended from the original condition and how best to prepare children to participate in caring relations. There seems to be an elementary form of sympathy in infants; the crying of one often produces crying in others, but this sympathy is very like that between two objects—vibrations in one set off harmonic vibrations in the other. Toddlers, however, seem to exhibit genuine human sympathy, often patting a crying child to offer comfort.

It seems obvious that people with well-honed empathic skills will be more likely to establish and maintain caring relations than people without these skills. If what we have said so far is right, women may be more likely than men to take responsibility for the maintenance of those relations. But for now, I do not want to talk about *caregiving* and a long-term commitment to act as carer. Rather, I'm interested in how we promote the disposition I've labeled "being prepared to care." How do we guide, train, or educate for empathy? What we want as a result of our efforts are people who will care because they want to do so—that is, people who will regularly establish natural caring relations.

The first step in learning to care is learning what it means to be cared for. The temptation is to think that all infants who survive have been cared for and, therefore, know what it means to be cared for. In today's world, however, this conclusion is unjustified. It is possible to keep a child alive and even to love that child without establishing a caring relation as I've described it. Unfortunately, more than a few children start school with no clear notion of what it means to be cared for, and early childhood educators report that their first task often is to help such children learn what it means to be cared for.

Teachers engaged in this task listen—pay attention—to the children and allow their motive energy to flow toward the expressed needs of the children. Teachers as carers do not ignore assumed needs (the sort that appear in the syllabus). They are, after all, adults and have some sense of what all children need, but they do not press for the satisfaction of these needs before a caring relation is in place. For example, a teacher may believe that all children should achieve a measure of independence, but she may allow a child to stay close, even cling, if that is what the child seems to need at the time. When the child's responses show that the caring has been received, he or she is on the way to understanding what it means to be cared for, and education for caring can proceed to the next step.

A problem arises here for care theory as I've described it. I've asserted that the response of the cared-for completes the caring encounter or relation. What if the cared-for claims indeed to be cared for, even though an impartial observer using the criteria we've established would deny that caring is taking place? Such situations are not rare. Children told repeatedly that rejection of their expressed needs is in their best interest may actually believe that they are cared for. Indeed, they may feel guilty if they suggest otherwise even to themselves. Often desperately unhappy, these children grow into adults who may inflict the same treatment on their own children.[26]

Technically, this problem can be avoided simply by pointing to the requirements laid down for the carer, but there are at least two good reasons for not doing this. First, we want our theoretical thinking to stay in touch with reality. In real life, there are pathologies of caring, and we should acknowledge them. When we consider trying to block pathologies through theoretical revisions, we should recognize turning points and go back to reality. Second, a technical rationale may save the theory, but it does nothing to rescue the troubled victim of pseudocaring. Help has to come through another human being who is prepared to care and to help the victim gain a clearer idea of what it means to be cared for. Teachers today are frequently called upon to do this.

The carer models what it means to care in her everyday caring. Not only does she do this in her natural caring relation with the cared-for, but she also shows the cared-for how to respond empathically to other people and to household pets. For example, in working with her youngster, a mother may say, "Tippy looks hungry. What do you think? Should we fill his dish?" She may also point out how happy Tippy is when the dish is filled. In talking to her child this way, she is helping him learn how "to read" Tippy's expressions. As the child gets older and takes more responsibility, the mother may point out and discuss occasions when the child fails in his responsibility. "Tippy was hungry, so I fed him. Poor Tippy. Did you forget? He looked very sad."

Parents who take this approach are using what Hoffman calls "induction." Writing about a form of discipline designed to promote empathy, he says: "The type of discipline that can do this is induction, in which parents highlight the other's perspective, point up the other's distress, and make it clear that the *child's action caused it*"[27] (italics in original).

But there is more to the story. Some parents make the mistake of bringing a pet into the home to teach their child (let's call him Billy) responsibility. They do not really like the animal—may even regard it as a nuisance. Billy will probably detect this attitude in his parents and conclude that they are not moved by sympathy for the animal; the pet is merely an instrument by which Billy is to learn responsibility. Billy may be at least minimally capable of the cognitive apprehension that accompanies empathy; he may see that his mother is not moved by empathy. If that is the case, Billy may not be moved by empathy either, and caring for Tippy becomes a duty. Even in the relation between people and pets, natural caring is in many ways preferable to ethical caring. Certainly it brings much more pleasure.

Dialogue accompanies modeling. In constantly renewed episodes of caring, carers help those receiving care to grow in their capacity for empathy by learning to read others. In her discussion of shaping an acceptable child, Sara Ruddick writes: "Many mothers find that the central challenge of mothering lies in training a child to be the kind of person whom others accept and whom the mothers themselves can actively appreciate."[28] All sorts of issues arise in this work, and I will discuss many of them in chapter 7, but it helps if parents choose a framework within which to conduct their moral guidance, and caring provides such a framework.[29] Consider, for example, the task of guiding the young toward polite and acceptable manners. If we concentrate on the development of empathy, we can explain why caring for others leads us to follow some rules of etiquette. We don't talk with our mouths full of food, for example, because the sight may be disgusting to our companions. Similarly, we smile and say "thank you" when someone holds a door open for us. We encourage children to notice and begin to anticipate the effects of their behavior on others, and we talk about these things. Many rules of behavior are designed to keep us from hurting others, and because we don't want to hurt others, we don't even need the rules. Such rules simply encode what we would do at our empathetic best. But other rules are mere conventions. It may be useful to know how to set a table properly, and showing that we know this may enhance our social standing, but it is not a skill directed to the well-being of others. Wise parents and teachers engage in regular dialogue on these matters. Successful dialogue on conventions that matter morally and those that do not might reduce much frivolous rebellion among teenagers.

If our objective is to produce people who will be prepared to care, we should concentrate on the development of empathy, not on obedience to authority. This does not mean that obedience to authority is to be ignored, but it should not be foundational in moral teaching. In dialogue, again and again, we emphasize our relational condition. The effects of my conduct on others should be primary in my thinking—not what will happen to me if I disobey.

Patterns of social genderization may exacerbate male-female differences. Girls are more often directed to consider the feelings of others, whereas boys are often encouraged to think of their own social status. The difference is captured in contrasting questions: How do you think Grandma felt when you said that? versus, What do you suppose Grandma thought of you when you said that? Both questions are worth reflection and discussion, but the first should be primary in encouraging empathy.

Moral education at home and in schools often concentrates on the acquisition of virtues and/or moral reasoning. Both have something valuable to contribute to moral growth, but there is a fundamental difference in emphasis between these programs and programs aimed at increasing empathy. When we try to inculcate honesty, courage, obedience, or courtesy as personal virtues, attention is directed to the moral agent, the one who "possesses" the virtue. In contrast, when we try to promote empathy, attention is directed to others—to those who are affected by our actions. Similarly, when we ask students to grapple with moral dilemmas (usually fictional), their attention is directed to impersonal principles and their correct application. However, when the objective is empathy, attention is directed to cognitive apprehension and sympathy first, and then to the instrumental reasoning required to produce a mutually accept-able solution to the problem. I do not mean to condemn or demean programs in character education or moral reasoning but, rather, to point out their limitations.

Girls are, more often than boys, asked to serve in caregiving roles. As noted earlier, caregiving is not always accompanied by caring. However, caregiving may give rise to caring, especially if it is guided by someone with heightened empathic capacities. This is something to keep in mind when we offer (or mandate) service learning in our schools and colleges. Students may learn much in these programs, but if we want them to increase their capacity for empathy, we have to be sure that their supervisors are giving some attention to this possibility. Too often, high school students engage in some form of service in order to augment their transcripts and further their chances for college admission.

Again, it is worth noting and discussing gender differences both in the context of expectations for caregiving and in the traits of character admired. In a prominent early (1909) text published by the Character Development League, thirty-two virtues are discussed—each with several biographical examples and portraits.[30] In most of the accounts, not one woman appears. Under *sympathy*, however, seven of the twelve portraits are of women! As I noted earlier, even the academic switch from sym-pathy to empathy must be viewed with some suspicion. *Sympathy* was thought to be too soft, too feminine, whereas *empathy* was more cognitive, more masculine, and therefore respectable.

An approach to moral education that concentrates on the promotion of empathy necessarily involves *guilt* and how that should be handled. Hoffman includes a discussion of "inductions" that produce guilt in children who have hurt others.[31] We should feel bad—feel guilty—when we have hurt another. Such guilt is aroused by the sympathy we feel for the one hurt and the knowledge that our action caused the hurt. The direction of feeling is toward the one hurt, and thinking is directed to restitution. In contrast, shame (used by some moral educators) is directed at oneself. It may or may not be accompanied by the guilt which motivates restitution and reconciliation. Instead, it often motivates excuses, anger, and attempts to protect the self.

But even guilt can become somewhat distorted. From the perspective of care theory, our guilt should be for the hurt we have caused to a particular person or group of persons. Hoffman's view involves some ambiguity. On the one hand, he clearly emphasizes the child's growing capacity to feel empathic distress and the child's own guilt for causing the distress. On the other hand, he claims that "mature guilt is a far cry from the early manifestations of guilt and guilt-like behavior in childhood." And "a lot must be accomplished developmentally before one is capable of mature guilt over *violating the norm of considering others*" (my italics).[32] A question we'll continue to pursue is, Should our guilt be over violating a norm, or should it be, as in sensitive later childhood, over causing hurt to *this person* or *these persons?*

Guilt-induction must be handled carefully. We don't want kids to be swamped in guilt over disappointing their parents by not making all A's, forgetting to mow the grass, not measuring up on the ball field, and the like. And sometimes, kids need protection from guilt. Martin Buber writes of the importance of *confirmation*.[33] As I've interpreted it, confirmation is an act that assures another that we understand that a questionable act may have had a better motive. To confirm another, we need to know that other quite well, and we must be capable of empathic accuracy. We can then attribute the best possible motive consonant with reality. Thus we might say to a child who has cheated: "I know you wanted to make your father proud, but that was not the way to do it. You are a better person than that." Confirmation is an example of what Murdoch refers to as trying to "see lovingly." In a related approach, it is reported that Goethe once said: "When we treat man as he is, we make him worse than he is. When we treat him as if he already were what he potentially could be, we make him what he should be."

As long as natural caring functions as it should, no real moral effort is required. Great expenditures of physical, mental, and emotional effort may be demanded of carers who are trying to meet expressed needs adequately, but there is no need to summon the ethical "I must." The internal, automatic "I must" is adequate. However, natural caring sometimes breaks down, and in today's global environment there is no possibility of caring *for* everyone whose cry for help reaches us. In such cases, we must call on ethical caring.

THREE

Ethical Caring and Obligation

In this chapter, we'll discuss the move from natural caring to ethical caring. We'll also take a close look at caregiving—long defined as "woman's work." Carework is both the incubator of natural caring and a current site of contention. On one hand, it represents the set of activities in which females have developed an enhanced capacity for empathy. On the other, society's expectation that women will continue to do the lion's share of care labor is a legacy of the second source of female empathy—subordination.

Discussion of caregiving will lead naturally to one on moral obligation and the limits on such obligation. I have rejected the idea that we are today operating with "Stone Age brains," but it may well be that our brains have not yet fully adapted to the enormous numbers of people with whom we can communicate and whose troubles we hear about instantly. We are flooded with cries of protest and pain, and these cries induce feelings of both obligation and impotence. How does an ethic of care help us in coping with the problem of obligation?

Finally, we'll ask whether care ethics must be restricted to small, face-to-face groups. Some have suggested that at a certain point care must give way to justice. I don't think that is quite right; the approaches need not be mutually exclusive. I will suggest that we try to develop a care-driven approach to justice.

WHEN NATURAL CARING FAILS

Natural caring sometimes fails even in the settings where it is most at home—in families, among friends, in small communities. We can be overwhelmed with worries or work, the cared-for may be exasperatingly difficult, or we may be frustrated by conflicts among those for whom we care. We live in webs of care, and it is often difficult to disentangle the various demands on us so that we can respond to each one with care.

What do we do when the natural inclination to care deserts us? Moral rationalists advise us to think conscientiously about the principles that should govern our behavior. If, for example, one of several children is especially cantankerous, how should we respond? There may be a principle of fairness we can draw on that will help us to respond justly, if not lovingly. Or we might put aside this child's current extreme needs and justify our decision on the basis of what is best for the whole family or web of care. Care theory advises us to draw upon our own ethical ideal—one built over a lifetime of natural caring. In effect, we ask, How would I respond if I were at my best caring self? This seems to be what Murdoch's mother-in-law, M, did; that is, she drew on her ethical ideal to interact considerately with her daughter-in-law. But she did more than that. Uneasy with her conduct as guided by ethical caring, she turned to her ideal of caring in a desire to establish a relation of natural caring. We might say that she drew on ethical principles to guide her initial, entirely acceptable behavior. But why, then, was she not satisfied with that? After all, she was behaving correctly. And why did she decide to "look again" in an attempt to "see lovingly"?

There is a temptation to say that care ethics does define a principle—something like "Always act so as to establish or maintain caring relations." Michael Slote makes the interesting point that "the ethics of caring is actually in a less good position to deny deontology than is act-consequentialism."[1] By this, Slote means that care ethics allows less leeway to violate rules against killing, stealing, lying, and other moral rules than various forms of Utilitarian ethics. At the theoretical level, he seems to be right. For example, most care theorists would accept as an absolute "Never inflict unnecessary pain," and we have already seen that care ethics agrees with Kantians on the Trolley and Lifeboat dilemmas.

But something is not quite right here. The acts of people who live by an ethic of care are rightly described by the principles mentioned above, but carers do not usually *refer* to the principles in making moral decisions. The principles are more descriptive than prescriptive. A great deal more research is needed on this, but there are already studies showing that some people behave honestly, altruistically, and even heroically in direct response to the needs of others.[2] They act without the expectation of reward and without reference to a principle. Slote argues (rightly, I think) that empathy—basic to care ethics—actually can guide conduct and produce behavior that looks the same as that demanded by deontology.

We might say that such people have so completely internalized the principles that they are automatically directed by them. But this suggests that the principles were somehow out there in the external world and were then internalized, and this again seems to be not quite right. As Slote argues, deontology may be anchored in empathy, not the other way around. The development of empathy in individuals is more like an accelerated, miniature process of evolution. Through continual interactions that require reading others and feeling for them, an individual becomes empathic—prepared to care. In chapter 6, I will discuss sympathetic attention as a primary virtue in care ethics. When natural caring fails, followers of care ethics turn to their own ethical ideal for guidance. What would this best self do?

In this turn toward a best self, care ethics is very similar to virtue ethics. Virtue ethicists, like care ethicists, depend on something internal to the moral agent; in the case of virtue ethics, they depend on the agent's virtuous character. But where does virtue come from? Socrates wrestled with this question—arguing at first that virtue was neither inborn nor taught. But although he insisted that virtue could not be directly, explicitly taught, he had to acknowledge that teaching must have something to do with the development of virtue. The issue is alive today. Character educators often call for the explicit inculcation of virtues and describe a constellation of virtues as character. In contrast, care theorists see virtues as traits encouraged by an environment rich in empathy and empathic guidance. Like pragmatists, we more often refer to virtuous acts than to virtues as permanent possessions of a moral agent.

A well-developed ethical ideal of caring may be more reliable as a guide to moral decision making than a set of principles. We often say, half in jest, "Rules are made to be broken," and it is certainly true that there are all sorts of reasons for setting aside various rules. In many cases, we invoke one principle to override another. And *duty* itself, so eloquently described by Kant, has been horribly corrupted when detached from inner reason and applied to a social/political hierarchy. Persons guided by an ethical ideal of caring do not simply break a rule when they fail to respond with care; they break something in themselves. We'll say more about this in a discussion of empathic nausea and violence in chapter 8.

Emphasis on an ideal of caring built over a lifetime of caring interactions does not entail a claim that rules and principles are worthless. As Virginia Held has pointed out, so-called universal laws are useful in some domains, but they tend to reduce and sometimes even destroy relations of natural caring. People in caring relations, Held writes, "are not seeking primarily to further their own *individual* interests; their interests are intertwined with the persons they care for. Neither are they acting for the sake of *all others* or *humanity in general;* they seek instead to preserve or promote an actual human relation between themselves and *particular others.* Persons in caring relations are acting for self-and-other together."[3]

Margaret Urban Walker has also warned against the pervasive use of traditional moral "knowledge," contrasting it with moral *understanding.* Contrary to philosophers who depend on abstraction as a way to avoid error, Walker suggests that in many cases, "adequacy of moral understanding decreases as its form approaches generality through abstraction."[4] It is this decrease in moral understanding that we seek to prevent when we use ethical caring to restore natural caring.

There are, however, many situations in which it is a physical impossibility to establish natural caring. Held speaks of "domains" in which ethics of principle are appropriate. In such cases, natural caring has not failed; it just cannot be established. Then we have to decide whether a dramatic shift from care to some other ethical theory must take place. This move is half right. We have to approach global problems differently, but perhaps we can inject important ideas from care theory into theories of justice. We'll consider this possibility in the last section of the chapter.

Issues of both care and justice arise in the arena of caregiving. Because care and caregiving are so often conflated and because caregiving has long been central in the lives of women, we will look at that next.

CAREGIVING

Many feminist philosophers write of care as labor.[5] I have maintained a distinction between caring as the fundamental concept in the ethic of care and caregiving as the set of activities associated with an occupation or form of work (paid or unpaid). The difference is important because caregiving may proceed with or without caring, and caring—as it is developed in an ethic of care—is a moral way of life, one that guides personal interactions in every domain of activity.[6]

What, then, is the importance of caregiving in our discussions of caring? I spoke earlier of caregiving as the "incubator" of values and virtues associated with caring. From an evolutionary perspective, this is clearly true, but it is also true that a well-developed ethic of care has much to contribute to a mature, humane view of caregiving. We need to consider both.

In caregiving driven by maternal instinct, females are concerned with the survival of their infants. The natural caring that has evolved from instinct gives rise to the processes already described in the discussion of empathy: learning to read the other, sympathy (feeling with), motivational displacement, and the reevaluation of what is read. But natural caring in the context of mothering also calls forth virtues such as responsibility, kindness, gentleness, patience, unselfishness, and cleanliness. Over time, immersion in the tasks of child-rearing has led beyond instinct to maternal thinking. Sara Ruddick has analyzed three great demands of maternal care: preservation (survival), growth, and acceptability.[7] Mothers (or maternal thinkers) want to preserve the lives of their children, foster their growth, and shape them in ways that will make them acceptable to their larger community.

In responding to these demands, all sorts of mistakes can be made. The moral approach we call *caring* can help to avoid many of these mistakes. For example, mothers guided by care ethics will not often use punishment and reward to push children toward acceptability. Rather, they will concentrate on the development of empathy, and in matters that do not concern harm or benefit to others, they will listen attentively to their children and assist in meeting their expressed needs. Motivated to maintain relations of care with their children, they will also remain aware of, and direct their children's attention to, the whole web of care. When care ethics employs the expression "web of care," we are not thinking beyond the reach of actual contacts to what

Utilitarians call the "greatest good for the greatest number." We are trying to be sure that our decisions do not harm anyone in the web, and if our decision to help one member also helps others, so much the better. But we will not cause great harm to one (or a few) in order to promote some presumed good for many.

One of the most valuable contributions of an ethic of care to the occupation of caregiving is its concentration on the quality of the caring relation. When we provide care, we do something for another, but because we are contributing to a relation, we thereby do something for ourselves as well. If we treat children with patience and good humor, they are likely to respond agreeably. Our work, imbued with both satisfaction and pleasure, may become easier—less like labor. Much the same can be said about the work of caregiving in all of the "caring" professions.[8] The main bulwark against burnout in the caring professions is the all-important response of the cared-for in recognizing the efforts of the carer *as caring*.

Perhaps because women bear children and tend to the direct needs of families, they have been called upon to provide for the needs of others who are unable to help themselves. Indeed, before nursing became a professional occupation, to call for a nurse meant simply to call for a woman, and daughters were expected to answer that call even if they had to give up paid work to care for family members. When a family member was not available, another woman from the community might have been brought in. In describing the early days of nursing, Susan Reverby explains that "nurses" were expected to perform whatever tasks were usually assigned to women in the household.[9] In addition to caring for their patients, they might do laundry, cooking, and cleaning. The caregiving provided by these women (family members or hired help) really was labor—hard, poorly paid labor.

When we consider the labor involved in caregiving, we must return to the second source of women's increased capacity for empathy. To ensure protection for their young, females accepted a position of subordination to their male partners. This has been, at best, a mixed bargain. I'm reminded of a cartoon depicting a woman in housecoat and slippers seeing her well-dressed husband (with briefcase) to the door. She says, "I'm glad I don't have to go beating out of the house every day at 7:00AM." There are some women even today who welcome their subordination as a good bargain. But for most women, staying at home has involved unpaid labor of some sort from morning until night. The "caring trap" identified earlier has been a reality. However, it should be called the *caregiving trap*, and caring as described in care ethics has often made the work more bearable, sometimes even enjoyable.

A major point in this book is that the bargain made by the earliest human females has handed down a legacy that still bedevils us. In today's occupational world, women often earn less than men doing the same work, and if they have chosen to work in one of the caring professions, they earn considerably less than men with comparable levels of education who have chosen more "masculine" fields.

It has long seemed "natural" for women to work in occupations similar to homemaking and child-rearing—that is, in occupations that require caregiving. However, the subordination of women aggravates that tendency. First, it is subordination—not the nature of the work—that results in lower pay and scant occupational prestige. What could be more important to the health of a society than the care and education of its very young? And yet the closer a woman's work is to that long identified with mothering, the lower its worth in our society. This pattern is part of a larger system in which traits are genderized, and those associated with males are granted a higher value—provided they are exhibited by males.[10] If they appear in women, they are likely to bring criticism on the women who exhibit them. Certain acts and attitudes shown by a man may be called "assertive"; the same acts and attitudes exhibited by a woman may be labeled "bitchy." Similarly, women are praised for being nice, gentle, kind, and unselfish, but these traits are not always admired in men, and their value in employment is not high.

Second, the empathic capacities of women often lead women to consider the welfare of others over their own. Besides the long-standing fault of a society that pays more for "men's" work than for "women's," it seems that women do not speak up for themselves.[11] Oddly, this is not, as some critics have claimed, because women are poor negotiators. It turns out that many women are exceptionally good at negotiating—but they negoti-ate for others, not for themselves. This other-orientation in women presents a paradox. On the one hand, empathy and emphasis on relation lie at the foundation of care ethics; on the other, the subordination accompanying the growth of empathy has encouraged women to be complicit in their own oppression. Should women learn to assert themselves as men do, or should men learn to be less self-promoting? Probably we need to work on both ends of this problem, and we'll say more about it in chapter 4.

There should, of course, be concern about the status of women in the professional world, and attention should be given to the special needs of women such as those accompanying pregnancy and infant care.[12] But attention should also be given to the women who engage in caregiving for a living. Too often, in this arena, professional women do not exercise their natural empathic capacities when they deal with the women who care for their children and clean their houses. Women providing childcare for well-to-do women are frequently left outside the circle of care—poorly paid and unheard. The situation of women doing paid carework must be considered both nationally and globally, but from the perspective of an ethic of care, the individual women employing caregivers must also give special attention to the problem.[13] Do women become less sensitive to the plight of specific others as they themselves become more successful in the public world? This is an interesting moral question. How often do women retain the personal characteristics of carer in relation to the women they hire, and how often does their concern shift to "caring about" and the expectation that "these problems" should be handled by the society as a whole? If that is the case, can we create a care-driven conception of justice?

OBLIGATION AND ITS LIMITS

In natural caring, the "I must" comes to us as something to which we must respond. Usually the processes of empathy guide us from attention through response, but some episodes cause conflict. A cared-for may ask us to do something we regard as wrong. For example, Ms. A may be asked by her son, Bob, to write a short essay for him so that he can get on with his other homework. Ms. A can turn down her son's request, yet meet the criteria of caring and make the episode one of caring. But why would she regard it as wrong to write the essay for him?

Ms. A, like most of us, has grown up in a society that regards cheating, stealing, and breaking promises as wrong. She wants Bob to be acceptable in the society of which he is part. At bottom, she defends rules against cheating, stealing, and breaking promises by pointing to the harm such infractions cause in the whole web of care. She and Bob are not alone either as individuals or as a mother-son unit. In cases where the child's act may cause pain, Ms. A will help Bob to feel empathic distress. But in many cases where there is no infliction of physical or emotional harm, Ms. A must explain her refusal to aid or participate by providing an explanation rooted in the reluctance to damage others in the web of care. In responding this way, Ms. A is working to maintain the caring relation with Bob and also to encourage an attitude of caring and thoughtfulness in Bob. She is preparing him for life in a community of caring. (An important side note: At the university level, we rarely take the time to explain our aversion to plagiarism in these terms. Students should be helped to understand how the web of knowledge creation depends on intellectual honesty.)

In making moral decisions, those guided by an ethic of care refer to the likely effects on other persons and their relations to one another. The final arbiter of their decision is not a universalizable principle, nor the greatest good for the greatest (anonymous) number, nor even their own character. Care ethics is close to virtue ethics in its reliance on something within the moral agent that directs a response, but should we call this something "character"? It does not seem to be a collection of virtues that can be drawn upon. I have called it an "ethical ideal of caring"—a memory of how we have responded over time in our best encounters as carers and how others have cared for us. This "ideal" is a set of actual events remembered, not one of imagined, wished-for responses, but it includes acts of reflection and evaluation. We cannot simply wish that we had responded with more care or that we might do so in the future. We have to reflect seriously, analyze, and actively prepare ourselves to do better. Acts of analysis and evaluation are included in the ethical ideal. The more successful practice a person has had in caring and being cared for, the stronger that ideal will be. That is why it is so important for parents and teachers to help children develop a capacity for empathy, put it into practice, and continually evaluate their success in building caring relations.

There are, of course, virtues associated with caring, but we do not draw on them one by one, and I'm not sure we could name any traditional virtue as absolutely essential to caring. If attention is a virtue, that comes close. But as we saw earlier, there is something more basic than attention—an attitude that opens the self to the other. Weil's question, What are you going through? may be asked with the hope that not much will be demanded of us, but it is asked; we hold ourselves open to the address of the other. And the carer's response, "I am here," can be made cheerfully or grumpily, but it is made.

This moral approach to life—a way of address and response—is most obviously applicable to domains in which encounters are face-to-face. Responding as carer becomes more difficult as the circle of communication grows. Ann Morrow Lindbergh identified the problem: "The inter-relatedness of the world links us constantly with more people than our hearts can hold. . . . My life cannot implement in action the demands of all the people to whom my heart responds. . . . Our grandmothers, and even—with some scrambling—our mothers, lived in a circle small enough to let them implement in action most of the impulses of their hearts and minds."[14] Part of the problem Lindbergh described is real, but part is a result of the mistaken conception of caring that we identified earlier. Caring is *not* synonymous with caregiving. It is a way of encountering others. As individuals, we cannot *care for* people all over the world, or even in the next city; we do not encounter them directly. Indeed it might be the case that the closer we come to an abstract, universal "caring," the farther removed we are from responding to the particular other who stands before us. We risk substituting formulas and programs for direct, personal response. It is one unfortunate legacy of bureaucracy that we have become dependent on isolated experts and institutions that specialize in particular human concerns. For example, teachers tend to retreat from individual responsibility as moral educators when official programs of moral education are adopted; that "subject" is now officially covered. On college campuses, faculty members withhold help they would once have provided their students in loco parentis and leave it to designated counseling services. And family members sometimes deny financial help to relatives because, after all, there are welfare programs.

On the positive side, institutions and collectives of various kinds can operate effectively over large distances and with large groups of people in need. Although we can neither care for nor act as caregivers for all those who need caring attention, we can support groups and institutions that may establish conditions conducive to caring. Many of us do give regularly and generously to organizations that serve the poor, hungry, sick, elderly, young, and displaced. However, because we are not in a position to receive the response of those we are trying to help, we cannot claim to have established caring relations. We trust the organizations to which we have contributed to establish the conditions under which *caring-for* actually takes place. Often it is hard to know whether our trust is well placed.

What is the role of individual moral agents in *caring-about?* Do we have a moral obligation to care about those we may never encounter, and if so, in what does this obligation consist? Peter Singer has called upon individuals in affluent societies to take some responsibility for relieving poverty in less affluent parts of the world. It is worth quoting him at some length:

> We could therefore propose, as a public policy likely to produce good consequences, that anyone who has enough money to spend on the luxuries and frivolities so common in affluent societies should give at least 1 cent in every dollar of their income to those who have trouble getting enough to eat, clean water to drink, shelter from the elements, and basic health care. Those who do not meet this standard should be seen as failing to meet their fair share of global responsibility, and therefore as doing something that is seriously morally wrong. This is the minimum, not the optimal, donation.[15]

On one level, this sounds reasonable, but problems arise immediately. If there is a moral responsibility, as Singer contends, is it an individual or collective one? Since we, as individuals, are unable to establish the caring relations that should complete our efforts, why make it a matter of individual moral obligation? If we extend Singer's thinking to other serious social problems, we find ourselves in an impossible situation.

Let's suppose first that my contribution will actually save a child from starving in, say, Africa. Is that the end of my obligation? Children need clothing, shelter, adult guidance, and education. They need, in short, not only to survive but to grow as individuals and to become contributing citizens. The concepts of "holding" and "staying with" are central in maternal thinking and in care ethics.[16] In care ethics, our obligation rarely ends with a justified decision or act. Life goes on after the decision, and we must tend to the relations we have established. Singer introduces this set of problems, but he does nothing further with them.[17] By recognizing this issue, I do not mean to suggest that it is therefore acceptable to "let them die" now rather than survive if that survival will lead to a life of poverty and misery. No. I raise the point as a possible challenge to individual obligation. There should be a better way of tackling these problems.

Consider what individuals may have to grapple with. If I hear the cries of children thousands of miles away, should I not also hear those of children in a nearby city who are living in shelters or inadequate housing, underfed and malnourished, breathing unclean air that has caused a near-epidemic of asthma, deprived of reliable parental guidance, attending unsafe schools? And what responsibility do I have for our overcrowded prisons and the continuous stream of new prisoners coming from the population just mentioned? Should I do something for the lonely elderly who fade away in some of our least attractive nursing homes? Shouldn't I expend some energy on convincing my fellow citizens that it is wrong for any person working full time at an honest job to receive poverty wages? What should I do for the homeless in the city across the river? What should I do about capital punishment, which I believe is wrong? And while I am contributing money to save the starving children in Darfur, what will I do when I hear about starving children in Haiti or about Cambodian girls forced into prostitution? As Lindbergh so poignantly put it, I cannot satisfy the demands of all those to whom my heart responds.

While we are worrying about global obligations, we surely have to accept obligations close at hand. As parents, we have obligations to our own children that we do not have for others and that other parents do not have for ours, and we recognize special obligations to immediate neighbors. If I am asked for money, it somehow matters whether the one who asks knows me and asks *me* directly or whether the one asking is a stranger making a plea to every passerby. I am not arguing that it is *right* to distinguish between someone we know personally and a stranger (although it might be), but I am arguing that this is the way we are constituted as biological organisms.[18] Singer and others try to convince us that biological or geographical distance should not affect our decisions (a life is a life). But a morality that flies in the face of biological facts will be hard to enact. Somehow we have to find a way for good people to meet the special obligations induced by biology and proximity and still do what they can to improve the lives of distant others. This will require collective action.

Both Singer and Peter Unger argue that distance is not (or should not be) a central factor in our moral decisions to provide help.[19] But as Michael Slote points out, distance has a dramatic effect on empathy.[20] Our sympathy is aroused more keenly when we are face-to-face with someone suffering fear or pain. This is a familiar phenomenon in combat, and we'll revisit it in chapter 8. It is much easier to drop bombs from a great height than it is to shoot someone point blank or stick a bayonet in someone. Face-to-face, we experience empathic distress, and that leads to motivational displacement. We feel the internal "I must," and we are compelled to help or to desist in hurting. One can argue that our feelings—however well rooted they may be in evolutionary biology—do not constitute a moral reason for caring more reliably in proximity. But a theory that insists on moral justification for our decisions must still acknowledge the power of moral motivation. And at bottom, our motivation may provide the only real justification for morality.

In care ethics, the cared-for contributes to a caring relation by somehow recognizing the efforts of the carer *as caring*. This recognition need not be gratitude, but it registers on the carer and completes the relation. Empathy is involved here too. The cared-for displays some empathy—at least cognitive apprehension—in recognizing the caring. And the carer who has exercised empathy in responding to the needs of the cared-for is eager to "read" recognition in the reaction. The empathic exchange of responses is essential to the caring relation. When Singer and others insist that distance doesn't matter, they overlook the fact that *completion* does matter, and distance often makes completion impossible. It is not *mere* distance, however, because we can complete caring episodes with those close to us in affection by letter, telephone, or e-mail. It is harder, and sometimes we long for the empathic response that can be seen or felt, but it is not impossible to maintain caring relations across great distances; however, it is very difficult to *establish* them across great distances. Recognizing the need of carers to receive some positive response from the carer-for, some charities devoted to helping children supply contributors with pictures of the children they are helping and, sometimes, with direct communication through letters. The bottom line here is that there is no caring-for without the completion that arises in encounter, and the caring-about that Singer so easily equates with caring-for is a different phenomenon.

Michael Slote has argued that moral agents should practice "balanced caring"—a conscientiously considered way of balancing caring-for those in our inner circle and caring-about the well-being of those we cannot actually encounter.[21] I think it may be possible to achieve a balance within caring-about, but it becomes more difficult when we try to include caring-for. Within caring-about, Ms. A describes the flood of requests to which she must respond: CARE, Habitat for Humanity, the Global Fund for Women, Wildlife Federation, Defenders of Wildlife, the Pearl Buck Foundation, and the Salvation Army; Amnesty International, the ACLU, the Innocence Project, the Women's International League for Peace and Freedom, and NARAL; several colleges and universities; and various local charities. In addition, some families make contributions to religious institutions, fraternities, and various groups representing medical services and law enforcement. All of these contributions must, of necessity, be somehow balanced, and Singer may not fully understand how conscientious many families are in their dedication to caring-about. Americans are much more generous proportionally than their government.

Although Ms. A and others employ a budgetary form of balance in planning contributions to satisfy their felt obligation to care-about, that planning may be disrupted. In a given year, because of an earthquake, tsunami, tornado, forest fire, rebel uprising, or epidemic, one organization may need much more than another. Or there may be a set of local emergencies: water damage in the library, vandalism at the local zoo, a family hard hit by catastrophic illness. Balance is tempered by the need to respond to emergencies.

But, aside from insisting that some reasonable amount be set aside for caring-about (when it is possible to do this), it is extremely difficult to balance caring-for with caring-about. Within the domain of caring-for, all sorts of emergencies arise—a child may need orthopedic footwear, an elderly parent may need a mechanical device for mobility, the family dog may have an accident and require an operation. These are emergencies that only the involved family can address. The easy mathematical equation that ignores relatedness, proximity, and the possibility of completion simply disappears. We have special obligations that cannot be ignored. The hundreds of dollars spent on the dog's operation might indeed save one or more children from starvation, but the situations are simply not comparable. This dog is *my* dog, and I took personal responsibility for her when I brought her into my home.

Another difficulty arises in trying to force a mathematical equation across cultures. Although few children in the United States face actual starvation, a substantial number suffer relative privation. There are reports now of impoverished teenagers refusing their school's subsidized lunches. They would rather go hungry than expose their poverty to classmates. A thoughtful community could eliminate this problem simply by providing free meals for all schoolchildren as part of the school day. It could, if necessary, solicit contributions from parents who can afford to make them. We should be concerned about the whole child, not only the child's physical hunger. We should address needs more holistically, and that would be one purpose of developing a care-driven approach to justice.

Before turning to that topic, let's return briefly to Singer's recommendation that people in affluent societies should give at least 1 percent of their income "to those who have trouble getting enough to eat, clean water to drink, shelter from the elements, and basic health care."[22] Many Americans already give much more than this. But the demands are great and unceasing, and we have to care for those in our inner circle. In this domain, we are faced with uncertainty, with all sorts of contingencies. Singer says that people who do not give the required 1 percent "should be seen as failing to meet their fair share of global responsibility, and therefore as doing something that is seriously morally wrong." This judgment is almost certainly unwarranted. Slote also argues that we have a moral obligation "to help strangers and people we only know about."[23] But he is more aware than Singer of the difficulties involved, and he does not mandate a specific (or minimum) amount to be given. In keeping with this, we need to know much more about the individuals who seem to fail in meeting Singer's requirement. What *are* they doing to improve conditions around them? What are they faced with in their family and personal lives? What are they going through?

To press his argument, Singer points out (rightly) that we would regard a person who walks away from a drowning child as morally reprehensible. What sort of person would walk by and let a child drown? Drawing an unconvincing parallel, he argues that failing to make a contribution that would save the life of a starving child is equally reprehensible. The person who fails to do this has done something "that is seriously morally wrong." But the situations are not comparable. If I rescue a drowning child, I *know* that I have saved *this* child: I have held her, her hands have grasped mine, her eyes have sought mine—first in panic, then in relief. I may lie awake that night wondering if the child (who now has an identity) will need medical care. Will her parents need help with the bills? How much can I afford to give? Moreover, and most obviously, because of the accident of my presence, I may have been the only one who could have saved this child.

In contrast, when I make a contribution to provide food for starving children far away, I can only hope I've saved a child. Has food reached the most needy? Or have bandits stolen it? Has some warlord seized the money and used it to buy guns? Are the aid workers safe in distributing the food, or are their lives put at risk? Singer may regard these questions as a set of excuses for failing to contribute. I do not offer them as excuses. Rather, they are offered to demonstrate the crucial difference between individual and collective obligation. I cannot morally reject significant needs right before my eyes. The internal "I must" clamors for a positive response of some sort. But Lindbergh was clearly right when she said that our lives "cannot implement in action the demands of all the people to whom my heart responds." It is physically and emotionally impossible for us as individuals to care-for everyone.

It may be that the real problems of obligation cannot be settled by philosophy. The discussions are instructive and help us to think more carefully, but in real life, we have to sort through a myriad of complexities. Perhaps we can come a little closer to a theory of care that reaches out to a wider web of care by considering a care-driven approach to justice.

TOWARD A CARE-DRIVEN APPROACH TO JUSTICE

It may be a mistake to insist that individuals have a moral obligation to give some of their income to relieve poverty in faraway parts of the world. To do so is certainly morally admirable; it is something to be encouraged, but it cannot be obligatory. Yet we feel the tug—a vague, somewhat hopeless "I must." Although we cannot care-for people we will never encounter, we can care-about them. Many, perhaps most of us, do care. But the complexities overwhelm us. Some recent research has shown that college-age students show less concern about social justice after a course concentrating on it than students who do not experience such a course. Why? One suggested answer is that students studying poverty and its woes begin to fear for their own futures—that they may suffer economic misfortune. I think it is more likely that they can't cope with the enormity of the problem. Many of them might say, in effect, Just take my 1 percent and free me to work on the problems in my own family and community. They may be onto something. One way to accept their generosity and free them is to recognize the collective responsibility of nations and to promote a care-driven concept of justice.

What would a care-driven approach to justice look like, and how would it differ from other conceptions of justice? The first point to be made is that there are several theories of justice, and the concept differs not only across cultures but, to an important degree, within cultures.[24] A care-driven approach recognizes these differences and does not attempt to universalize any one version. We start with a somewhat vague notion that people committed to justice are dedicated to doing right by everyone.

Care and justice are often contrasted, and a substantial literature describing the contrast has emerged.[25] Virginia Held captures the contrast as it is popularly construed: "An ethic of justice focuses on questions of fairness, equality, individual rights, abstract principles, and the consistent application of them. An ethic of care focuses on attentiveness, trust, responsiveness to need, narrative nuance, and cultivating caring relations."[26] Sometimes, theorists locate the salient differences in domains of application—care properly dominant in face-to-face interactions or relations and justice dominant in larger public domains. Even those who take this position—as I did in earlier work—acknowledge that the two ways of thinking overlap across domains. We may want justice in the sense described by Held, but we want it tempered by care. And as already noted, care theorists speak often of "balanced caring," by which they seem to mean caring incorporated in a larger system of justice.

In *Starting at Home* and in the preface to the second edition of *Caring*, I suggested that *caring-about* may be thought of as the motivational foundation for justice. This means, of course, that *caring-for* remains basic in our thinking about justice (since it underlies *caring-about*) and, more generally, about morality. I do not claim, however, that *caring-for* is the only conceptual starting point for a theory of justice. Clearly, one could make a good case for self-interest and prudence as viable starting points, and I do not want to claim that caring necessarily underlies every conception of justice.[27] Rather, I am interested in developing an approach to justice that does build on fundamental concepts of care.

We properly care-about the needs and sufferings of many people whom we are unlikely to meet face-to-face. We hear about their needs and wish we could help. In contrast to Kantian theories of justice in which *rights* play a major role, *needs* are basic to a care-driven approach. Caring is itself a needs-based ethic. Chapter 7 is devoted to a discussion of needs and wants. We might use the word *justice* to name a system through which our caring-about responds effectively to the needs of others. A first step is to establish lines of communication—to listen attentively to the needs expressed and to learn something about how these people conceive of justice. It should be an assigned task of every overseas agency to maintain and strengthen such lines of communication. If we as a nation plan to conduct affairs of any sort in another country or culture, we should be sure that the people of this other nation are involved in the plans for their future. The economist Joseph Stiglitz makes this point at the very beginning of his discussion of globalization: "Those whose lives will be affected by the decisions about how globalization is managed have a right to participate in that debate, and they have a right to know how such decisions have been made in the past."[28] I would prefer to use *need* rather than *right* in this context, because *right* already suggests a conception of justice that may not be embraced by the other group. Our task is to work together to forge a concept of justice that both can accept, or to reach an agreement that allows the groups to hold different views of justice without breaking off dialogue.

There are certainly views that we (for now let's restrict the "we" to representatives of Western liberal democracies) cannot endorse. We cannot endorse slavery as consonant with justice, nor can we accept official discrimination against women. We are uncomfortable with officially sanctioned female genital mutilation. It is not possible that a collaboratively defined concept of justice should include practices that are anathema within our own. A shared concept of justice may be put on hold—discussed but not ratified. We can, however, reject the temptation to impose our own views—to apply sanctions or other coercive measures to convert others. Instead, we should persist in dialogue, explain continuously why we find some practices unjust, and increase opportunities for ordinary people in both countries to become acquainted and work together on common projects.

In a care-driven system of justice, we would be unlikely to isolate those who disagree with us on the meaning of justice. Notice that Stiglitz's advice on including in conversation all those whose interests are at stake is compatible with Western views of justice. We believe in both rights and participation. But we also must deal with people whose views are very different and even with those whose practices violate our sense of justice. The response most often invoked is to isolate the offender, withdraw our own citizens, and apply sanctions of some sort. An alternative, suggested by care-driven justice, is to invite more visitors from the offending country and send more of our own citizens to live and work in their land. At the present time, for example, we should invite more Iranian students to study in the United States, and we should send more of our students to study in Iran. We should also increase cultural exchanges at every level— in the arts, crafts, building industry, medicine, education, and every other walk of life where common interests can be identified.[29] The idea is to saturate the other with our presence, to establish relations of care and trust as part of preparation for diplomatic negotiations aimed at reconciling our difficult political differences.

E. O. Wilson's recent short book, *The Creation*, provides one model for such an approach.[30] Wilson casts his book as a letter from a secular humanist (Wilson himself) to a southern Baptist pastor. He honestly lays out their religious differences at the outset. But he then suggests that they can put aside these differences and join together to save the creation, to avoid environmental collapse. In addition to confessing their differences in plain terms, Wilson also points out areas of agreement. He writes, "My guess is that you and I are about equally ethical, patriotic, and altruistic," and he closes his letter with these words: "In closing this letter, I hope you will not have taken offense when I spoke of ascending to Nature instead of ascending away from it. It would give me deep satisfaction to find that expression as I have explained it compatible with your own opposing beliefs. For however the tensions eventually play out between our opposing worldviews, however science and religion wax and wane in the minds of men, there remains the earthborn, yet transcendental, obligation we are both morally bound to share."[31] Wilson's invitation to set aside significant differences and work together on common projects is illustrative of the approach taken by a care-driven concept of justice.

Wealthy nations, bonding together, could accomplish much to improve the conditions for people living in impoverished nations. Besides insisting that every agency involved should engage in dialogue and encourage cooperative activity, they should establish a coordinating agency that would assess conditions throughout the nations to be helped and evaluate the success of their combined efforts. At present, there are organizations whose purpose is to solve one great problem, such as the elimination of AIDS or malaria. This seems entirely appropriate for certain, specific organizations. But it is not appropriate as an overall approach; we need a more holistic approach. A coordinating group must look at the entire web of care and see how various problems impinge on the lives of the people affected. It is good to eliminate disease by vaccination, for example, but such efforts should be accompanied by improved living conditions—clean water, adequate food, universal elementary schooling, and all those features of life that contribute to the sustainability of individual improvements. Earlier, I gave a very simple example of the kind of thinking necessary. When we provide free lunches in our high schools for students "eligible" for such charity, we should think also of how to protect students from the stigma of accepting our charity. At home or overseas, we are dealing with whole people, complex situations, and an entire web of care.

In the approach to justice advocated here, we do not deny our differences, and we certainly do not shrug off abhorrent practices with "it's just their way." But we pursue common values and shared projects not only to accomplish important ends but also to know one another better and to set the stage for frank discussions that may lead to the abandonment of practices we find unjust. We hope that, through closer contact and cooperative activity, groups and nations that might otherwise be shunned may recognize the values of liberty, rights, and participatory democracy. But *we* may also come to appreciate values we now reject or misunderstand. A care-driven approach to justice will rarely authorize coercion. It will operate by establishing the conditions under which caring relations can flourish.[32] In most cases, the maintenance of caring relations is accomplished by mutual sensitivity to needs and conscientious attempts to meet them.

It might be even better to say that a nation, culture, or organization should locate and support circles of natural caring rather than to charge the helping group with establishing such circles. Circles of natural caring exist in all cultures, and provided they are not dedicated to the extermination of other groups, the task of an intervening organization is to support these groups and provide communicative links among them. The task is *not* to convert others ideologically to the position of the helper or "liberator." In working with large groups—at the national or international level—our purpose should be to coordinate and expand circles of care by providing chains or links of connections. Ideological differences may never be overcome, but the desire of one to eliminate the other should disappear in the recognition of multiple common interests.

So far, in taking a care-driven approach to justice, the emphasis has been on collective responsibility. It is nations and groups organized across nations that must take responsibility for social/economic problems at the global level. But what is the connection between individual and collective responsibility? Individuals can participate in the cross-cultural groups discussed above. Few of us would want to insist, however, that they are morally obligated to do so. In discussing what affluent people owe to the earth's poor, Thomas Pogge writes that it is our duty "not to collaborate in upholding an institutional order that avoidably restricts the freedom of some so as to render their access to basic necessities insecure without compensating by protecting its victims or working for its reform."[33] A statement like this provides months (perhaps years) of work for industrious philosophers, but it gives little moral guidance to reasonably intelligent citizens who are trying to care-about people for whom nations and other collectives must take responsibility. What does it mean to refuse collaboration in upholding an institutional order? How am I to do this? As a citizen, I have no real control over how my taxes are used. If I did, we would never engage in preemptive war. Does it mean to vote appropriately? But candidates for office regularly say one thing before taking office and something quite different when they have been elected. Does it mean to speak out—to express myself forcefully on the topic? How can I be sure that a particular "institutional order" is the one at fault? Does it mean that I must quit my job if the company for which I work is in any way involved with the offending institutional order? All of these questions are worth exploring at some length, but it is not easy to see how that exploration can culminate in a statement of moral obligation. Yet I agree that we have some obligation, the nature of which must be explored. Again, I think that many conscientious citizens, given the choice, would voluntarily check a box on their tax form that invites them to contribute 1 percent additional tax toward the elimination of world poverty.

If we contrast Pogge's vague statement of negative duty with the obligations staring us in the face, we are likely (and perhaps rightly) to return, with Candide, to the cultivation of our gardens—to ensure decent pay for the women who care for our children, to protect the emotional as well as physical health of our schoolchildren, to assist our neighbors and, yes, to attend to the expressed needs of family members.

Beyond the incessant demands that bring both predictable and unpredictable obligation, we have some obligation to endorse, perhaps to promote, a care-driven approach to justice, but I would not dare to say exactly what each and every person must do. In this brief introduction to a care-driven approach to justice, the following points seem essential:

Circles of natural caring should be identified and respected in each culture. Governing bodies, cooperatively chosen, should serve the function of linking these circles for the accomplishment of cooperative ends.

Collective groups—national or international groups— should work collaboratively to define a form of justice that meets the legitimate needs of all participants. The emphasis is on *needs*. The satisfaction, or at least recognition, of needs is essential in maintaining caring relations.

Groups should maintain dialogue in the face of basic disagreements. The dialogue should not exclude differences, but it should put them aside when relations of care and trust are at risk. Talk about something else; work collaboratively on some mutually chosen project.

In times of conflict, groups should increase contact with one another. Isolation and sanctions should be rejected. The idea is to form relations of care and trust. When it has become unthinkable to do physical harm to one another, that is the time to discuss what seem to be irresolvable differences.

Groups should identify projects in which opposing parties can cooperate. Although fundamental differences remain (in, for example, political or religious values), they can work together to achieve common goals.

Projects aimed at the solution of one great problem—for example, eradication of malaria—should be subsumed in holistic programs that consider the whole lives of individuals and societies.

We will return to details in the development of care ethics in chapters 5 and 6, but first we need to explore a concept that seems to separate care theorists from feminists who embrace liberal theory. That topic is autonomy.

FOUR

The Limits of Autonomy

In this chapter, I will look at autonomy from several angles. First, I'll consider the familiar social/political idea of autonomy as individual independence, or self-sufficiency. Second, I will spend some time on the influential Kantian concept of the autonomous will and some significant critiques of that concept. Third, having rejected important facets of both previously discussed concepts of autonomy, I'll explore a possibility compatible with care ethics—that of limited autonomy conceived as choice and responsibility within a certain span of control. Finally, I'll discuss the connection of critical thinking to what might be called intelligent heteronomy.

AUTONOMY AS INDIVIDUAL INDEPENDENCE

Martha Fineman writes of "the autonomy myth" in an attempt to construct a persuasive theory of dependency.[1] When we say that an individual, group, or nation is *autonomous*, we usually mean that it is not under the rule or control of other individuals, groups, or nations. In a political context, we often identify autonomy with freedom. In a social/economic context, we see it as almost synonymous with self-sufficiency. An autonomous individual can take care of herself or himself. In the United States today, we put a high premium on the autonomy of our nation and its individual citizens. Fineman comments: "Our all-American hero is therefore the autonomous individual, protected by law from unwarranted interference with his rights by other individuals and by government on any level, and free to conquer the frontier, be it westbound or upward into space. The rhetoric of individual freedom and rights incorporating an ideally restrained and limited government permeates our society."[2]

In calling autonomy a "myth," Fineman is not labeling it a falsehood, nor is she rejecting the idea of autonomy entirely. In the United States, we have elevated the concept to a myth. Myths are enormously powerful; although subject to revisions and distortions over generations, they claim continuing devotion. To reject a national myth is to risk having one's national allegiance or patriotism called into question. Powerful myths infect whole cultures, and it is not surprising that Americans who cherish their national autonomy also admire individual autonomy and look with pity or contempt on those who are not self-sufficient.

An ethic of care attacks individual autonomy at its roots. It agrees with Fineman that we are all born dependent and that many of us require "care" at various times in our lives. Fineman uses *care* as I've been using *caregiving* or *caretaking*, and she is surely right to point out the continuing human need for care in this sense. However, care theory goes further and insists that relation is ontologically basic and the caring relation morally basic. We become individuals only within relations. We are recognizable individuals as separate physical entities, but the attributes that we exhibit as individuals are products of the relations into which we are cast.

Most of us desire and claim a certain amount of freedom in organizing our own lives. This freedom has come relatively late to women, who until recently have been expected to recognize men as their masters. Even today, some religious groups in the United States insist on a dominant role for husbands and for males generally. I have suggested that the long centuries of subordination forced women to learn to "read" the males who directed their lives. The resulting capacity for empathy should be highly valued, but subordination should end.

Subordination and dependence are both looked down upon in a society that has embraced autonomy as a myth. All those tasks that have been assigned (or "fallen naturally") to women have been undervalued and underpaid. It is not that caregiving tasks are unimportant. Most reasonable people admit that the quality of care provided for children, the disabled, and the elderly is a mark of a society's goodness and decency. The tasks are undervalued because they have been performed by women, and women have been considered inferior to men. It follows that tasks performed by women must be inferior tasks.

In seeking a "tenable state," Fineman recommends that the importance of caregiving be recognized by law. As a foundational commitment, the society must take collective responsibility for dependency. A progressive, democratic state "would provide two different types of fundamental social goods—

[including] housing, health care, a minimum income guarantee, and other necessities. . . . The second type of subsidy, which is specifically directed at supporting caretaking, requires the state to ensure both material and structural accommodation."[3] A move toward the first sort of subsidy in the United States will almost certainly occur, but it will be slow and highly contested. While much of the Western world has made substantial moves in this direction, Americans are still horrified by anything resembling socialism. "Socialized" medicine or anything "socialized" is anathema to many Americans.

I want to spend a bit of time here on the second type of subsidy and its correlates in education and public discussion. In agreement with Fineman, I think government must be involved in regulating the practices of employers with respect to the caregiving responsibilities of caregivers. There should be a way to reduce the professional penalties women pay for acting as caregivers.

Illustrative of the political attitude toward caregiving is the current interest in limiting or even canceling the nonprofit, tax-free status of some childcare agencies. In some cases, the agencies in question do charge for their services, but so do many other nonprofits. It even seems reasonable to challenge the tax-free status of enormously rich universities and evangelical churches, but authorities are challenging childcare agencies with the question, Since you charge, what is it that you are *giving?* People who ask this question fail to notice that some medical and eldercare facilities operate as profit-making businesses but are nevertheless subsidized by government programs such as Medicare. Prekindergarten programs are sometimes available, but they do not provide infant care, nor do they address the needs of parents whose children are occasionally ill and unable to attend school. It is ridiculous to suppose that charitable agencies could provide high-quality, affordable childcare without considerable financial support from government. The question, What are you giving? is indicative of the low status of children, mothers, and childcare workers.

The problem is first a social problem, one of status and ascribed worth. Working women want affordable childcare, but subsidized care is sometimes of poor quality, and it does not meet the need to care for children who are ill, for those over the age of six, or for those whose parents work odd hours.[4] Like the general public, many professional working mothers are unwilling to pay a respectable wage for childcare. Instead, they often hire illegal immigrants, pay a low wage, and avoid paying social security and other benefits. (Recall the argument I advanced on this problem in the previous chapter.) Thus, a widespread attitude neighboring on contempt is directed at childcare and its workers. It is astonishing that so many citizens—both women and men—do not see the contradictions in the positions they take toward childcare. They seem to have adopted what Orwell called *doublethink*—"the power of holding two contradictory beliefs in one's mind simultaneously, and accepting both of them."[5] One belief holds that children are our treasure and our national future; the second that their caregivers are not worth much. We want affordable childcare, but we do not seem to care if the workers who provide it live in poverty.

An ethic of care builds upon our desire to respond positively to need. When we cannot do this as individuals, we must draw on a care-driven concept of justice. But the collective will to build such a concept and live by it depends on a dramatic change in social attitudes. This is, at least in part, an educational problem, and here we are up against the long-standing complicity of women in their own denigration.

At the time of this writing, we hear numerous commencement addresses. Most call eloquently for graduates to commit themselves to public service of some sort for at least some time. "Ask not what your country can do for you but what you can do for your country." Graduates are urged to join the Peace Corps, to become community organizers, to serve in the military, or to volunteer in some political activity. They are encouraged to do something to make our country and the world a better place, not simply to make money. But the message that comes through (although hotly denied) is that engaging in this sort of work for a few years will bring personal, not just national, rewards in the long run.

How often have you heard commencement speakers urge graduates to engage in childcare? Where in our schools do you see any commitment to serious learning about and involvement in parenting? Is there nothing to be learned in this area? Can you imagine a commencement speaker advising graduates to spend two or three years with young children? Or to become a nurse's aide and learn something about suffering and the hard work of alleviating it? In our "best" high schools today, young women are advised away from the caring professions. A bright young woman who wants to be an elementary school teacher is likely to be told that she is "too smart for that"; she should set her sights "higher." And most women, subordinate even in our thinking, agree.

There is surely a set of educational problems here, but women too often work against reasonable solutions. Women should, of course, have access to the occupations that have conferred status and wealth on men. We should have some control over our own lives and futures. But what of the activities for which we have had responsibility for centuries? Should we agree with powerful men that these occupations, paid or unpaid, are worth very little? The discussion here leaves us with an uneasy feeling that, although we want to control our own lives, we may be unavoidably heteronymous in our thinking. That worry raises a deeper question concerning the notion of autonomy. In what sense, if any, are human beings autonomous?

THE POSSIBILITY OF AUTONOMY

Philosophers and theologians have argued for centuries about the problem of human free will versus determinism. Either position, taken as absolute, leads to complications that may be irresolvable. I argued earlier for the value of turning points in our thinking and theory building, points at which we should turn away from further development of abstract schemes and consider actual, concrete situations.

Determinism has grown in popularity as science has grown. It is thought by some that everything, including human behavior, has determinate causes. Clarence Darrow, the preeminent attorney who defended John Scopes in the evolution trial, held this view and sometimes used it with powerful effects, for example, in defending Nathan Leopold and Richard Loeb, wealthy young University of Chicago students who were charged with kidnapping and murdering fourteen-year-old Bobby Franks. Both charges carried the death penalty on conviction, and there was no question of the defendants' guilt. They had confessed, although they later entered a not-guilty plea, which, on Darrow's advice, was changed again to *guilty*. Darrow's only hope, in presenting an eloquent defense, was to save the boys (nineteen and eighteen years old) from hanging—no easy task in a climate of public opinion clamoring for the death penalty.

Darrow could not argue, as he often did, that poverty and neglect had shaped the young men to be criminals. Both Leopold and Loeb came from wealthy families, and they were students at a prestigious university. Darrow argued instead that the boys, although not technically insane, were "mentally diseased." Employing the premise of determinism, he argued that the highly intelligent defendants were without normal human emotion:

> I know that they cannot feel what you feel and what I feel;
> that they cannot feel the moral shocks which come to men
> that are educated and who have not been deprived of an
> emotional system or emotional feelings. I know it, and
> every person who has honestly studied this subject knows it
> as well. Is Dickie Loeb to blame because out of the infinite
> forces that conspired to form him, the infinite forces that
> were at work producing him ages before he was born, that
> because out of these infinite combinations he was born
> without it? If he is, then there should be a new definition
> for justice.[6]

Darrow emphasized that among the "infinite forces" that had formed Loeb and Leopold were elements of the boys' education; Leopold "had been corrupted by reading Nietzsche's philosophy, taught to him at the University of Chicago."[7] At this point, many of us would reject Darrow's argument on the grounds that countless people have read Nietzsche without becoming murderers. However, Darrow argued that, in combination with other infinite forces, the study of Nietzsche might have had some influence.

But there is a stronger reason to turn away from Darrow's argument and its underlying premise of determinism. What of the professors who included Nietzsche's work in their syllabi? Could they have chosen to omit Nietzsche? Most of us would say, of course they could have. They were influenced, partly shaped, by their own education and the power of philosophical thought at the time, but surely they could have chosen to omit Nietzsche. We could argue that their choices should have been supervised, that thoughtful censorship should be exercised in schooling at every level, but we would then be launched on another problem entirely. (One on which, interestingly, Darrow took the side opposing censorship.) However we feel about Nietzsche and censorship, we are led to observe a turning point. Convinced that professors have a genuine (if limited) choice, we turn away from absolute determinism. Indeed, the contemporary philosophical position is to reject both absolute determinism and absolute freedom, and that is the position I will take in what follows.

I am not claiming that we have demolished determinism. We could go on to analyze the concept of cause, to discuss the roles of emotion, motive, and desire. We could press an argument for the chemical causes of emotion. All of these possibilities are interesting philosophically. I am not alone, however, in recognizing turning points in these arguments. Our experience forces (?) us to believe that we have some control over our choices, that we are indeed moral agents. Because the vast majority of us believe this and our laws are framed as though it is true, we must work with the social consensus that we are indeed moral agents, responsible for our choices.

Darrow played successfully on the emotions of a judge who was not immune to emotional arguments. As we said earlier, empathy can sometimes make us dislike a person more as we come to know more about that person. If we are predisposed to vengeance or retributive justice, empathy may increase our antipathy. However, a person who usually keeps open the channel between knowing and feeling may experience sympathy with the increase in empathy. Darrow managed to move both judge and courtroom audience to tears of sympathy for his guilty clients.

In opposition to determinism, some philosophers have long argued that not only are we free, autonomous, but that such autonomy is an unavoidable precondition for fully human life. Kant was not the first thinker to argue for freedom of the will, but he is surely one of the most influential. We can certainly agree with him when he points to the fact that humans see themselves as having free will and that the idea is sure to be influential in guiding our actions and judgments. But he leaves the concept of free will unexplained, even as he posits a "good will" as fundamental to moral action. It is hard to reconcile the claim that we have free will with the insistence that as moral agents we are entirely governed by the "moral law within." We must freely give assent to this moral law, but once we have done so, are we still free? Even committed Kantians recognize that there is a problem with defining the moral subject and an autonomous will as preempirical—existing prior to any engagement with the real world.

Another important objection to Kant's position on morality arises in response to his insistence that the moral law is entirely a product of reason. This claim was, in part, a reaction against the sentimentalism of Francis Hutcheson and its opposing claim that morality is based in affect and emotion. With his claim for the exclusive reign of reason, and thus of duty, in the moral domain, Kant arbitrarily dismissed women as moral agents, because he believed that women did not possess the reasoning capacity to participate in genuine moral life. He acknowledged that women often do the right thing in many situations, but they do this out of a kindly nature—not as a result of reasoning from which they conclude that an act is one of duty. For Kant, there is no moral credit due for acts done out of love or inclination; he rejected the idea at the very heart of care ethics. Thus, the Kantian notion of an autonomous will subject only to reason is not compatible with an ethic of care. Care theory anchors itself in natural caring and uses the power of reason to maintain it.

More recently, existentialist philosophers have extended discussion of autonomy and human freedom. Jean-Paul Sartre, for example, made freedom the very basis of his description of human consciousness. For Sartre, our freedom is frighteningly complete and, on recognizing that freedom, we suffer anguish—sometimes to the point of nausea.[8] We cannot escape our freedom, although we can deny it and live in bad faith.

Viktor Frankl, an existentialist psychiatrist, also saw freedom as one of three factors characterizing human existence. A survivor of the Holocaust, Frankl saw clearly that we cannot always control what happens to us. Jews in Nazi-controlled Europe had no control over their physical conditions. However, Frankl said, they could still choose their attitude toward their suffering.[9] I have argued strongly against that claim.[10]

Consider what happened to Winston Smith in Orwell's *1984*. Imprisoned and tortured by the evil O'Brien, agent of Big Brother, faced with his greatest fear—rats—Smith ultimately betrayed Julia, the woman he loved. To save himself, he begged that the rats be removed from his face and set upon Julia. Something of the same sort happened to Julia. Both were morally destroyed. While he was still able to reflect on the matter (even that capacity faded away), Winston thought: "'They can't get inside you,' she had said. But they could get inside you. 'What happens to you here is *forever*,' O'Brien had said. That was a true word. There were things, your own acts, from which you could not recover. Something was killed in your breast; burnt out, cauterized out."[11] *Nineteen Eighty-Four* is fiction, but there are enough real-life cases to convince us that it rings true—that malevolent others can indeed destroy not just our lives but our moral identity. The human moral self is an empirical entity shaped by both genetic and social factors.

We have come to another turning point; we are not absolutely, completely free. We are at the mercy of things done to us, things that happen to us. The questions now must be, Given that we have *some* freedom, how shall we describe it? How do we acquire or lose it? How should we exercise it?

THE SPAN OF CONTROL

So far, I have rejected views that place either a real or metaphysical individual before its ends or desires or, even, before the influence of the groups into which she or he is born. Kant's transcendental, autonomous self—beautiful as it is—has no basis in reality. Similarly, Rawls's hypothetical thinker in the "original position" behind a veil of ignorance can be little more than a fiction—part of a game that ends with the intrusion of real life.[12] Some liberal feminists also subscribe to the priority of the individual over the group. Martha Nussbaum, for example, takes this position when she says that "the flourishing of human beings taken one by one is both analytically and normatively prior to the flourishing" of a group.[13] This is not to say that the theoretical results and recommendations achieved by these questionable beginnings are not admirable. They deserve thoughtful consideration but, perhaps, a different theoretical base. The liberal notion that distinct individuals precede the formation of relationships is contrary to what is easily observed in human life. The preempirical self is a Kantian ghost.

Some sense of autonomy—some concern with the control of one's own life—is vital to care theory. I have already discussed the difficulties faced by care theory when *caring* and *caregiving* are equated, and I have recognized the long-standing complicity of women in their own subordination. Indeed, the most damaging feminist objection to care theory is that it seems to endorse the self-sacrifice and subordination of women. It is, therefore, especially important for care theorists to suggest and elaborate upon a defensible view of autonomy.

Feminists have started to describe a form of *relational autonomy* that recognizes a relational self and is concerned with options, opportunities, and competence.[14] My preference would be to speak of limited control rather than autonomy, because of its long association with individualism and male dominance. However, it may be strategically wise for feminists to use a revised version of the concept (autonomy) that is so central to the myth of Western liberal democracy.

It is clear that human beings are not autonomous (free to choose) in many of the categories governing our lives. We do not choose our parents, the cultural groups into which we are born, our first language, our economic status, the genomic patterns that predict our physical characteristics and talents, or our first religion. On this last—religion—liberal philosophy sometimes exaggerates our opportunities for choice. It is far easier for a young person brought up in a lightly held religious connection to switch or drop religious affiliation than it is for one whose whole identity is tied up in religious origins. I have encountered more than one graduate student who has had to give up an entire community and all original associations—even family—in order to go to graduate school. The story often reduces both the teller and the audience to tears. The change can be made, but it is extraordinarily difficult.

Usually, when a decision is made to give up religion, family, or nationality, it is because the decision maker seeks greater control in those areas within the span of control. Most of us want to limit external control over our lives, choose an occupational path, and work out our own opinions on matters of importance to us. In the United States today, most of us feel sorry for women who are prevented from choosing their own marriage partners, educational preparation, and occupational life. But even in cultures that closely prescribe how women must live, there is a span of activity in which they exercise some control.

How do we decide what is or ought to be under our control? Oddly, the beginnings of whatever control we eventually achieve are heteronymous. If our culture is one in which the formal guidelines are anchored in liberal-democratic thought, we will generally have a wider span of possible control than those in more tightly ruled societies. If, within a liberal-democratic cultural, we are also free from religious hierarchical rule, we will have considerable independence. And if we are blessed with parents who encourage us to make wise decisions, we may achieve a substantial degree of autonomy. Even so, our span of control will tend to expand or contract as our situation changes.

In one basic way, because we are committed to care, we bind ourselves. Earlier, we discussed the paradox of Kant's autonomous self being tightly bound to the moral law within. How free are individuals whose decisions are so completely defined by the moral law that they are all bound to respond to a given situation in substantively the same way? But one may also ask about the effects of our being committed to respond to others with care. Our span of control is limited but yet substantial, mainly because the emphasis in care ethics is on caring *relations*, not on caring individuals. I have claimed that, in most cases, when we respond as carers, we do something for ourselves as well as for the recipient of care. By strengthening the relation, we help both parties in the relation. Still, as several sympathetic critics have pointed out, conflicts of interest will arise, and people committed to caring have to find a way to handle them.[15] Care theory suggests an approach to handling conflicts, but it does not bind us to Kantian-like prescriptions.

Although we are defined in relation, we are individuals— separate physical entities and different selves. I'll say more about the relational self in just a bit. But because conflicts of self-interest do arise, we must respond with sympathy and understanding. The guiding aim is to establish equal, mutual relations with other competent adults. In close, mature relationships, we expect to exchange positions regularly—acting as carer in one encounter, cared-for in another. In maintaining caring relations, we seek what Virginia Held calls "mutual autonomy" and what Martin Buber has called "equal relations."[16]

Conflicts of interest are, then, discussed, negotiated. Ideally, both parties have a well-developed capacity for empathy and can adjust to the needs of the other. If, in a relationship that should be one of mutuality, a woman has to serve as carer so often that she begins to feel resentment, she should also realize that she has not been strengthening the relation by her actions. Her efforts at caring have somehow misfired, and she must do something different—perhaps leave the relation empty of encounters for a while. "Caring" to the point of self-sacrifice often defeats its purpose. When it fails to maintain the relation as one of caring, the carer must take a new direction.

So far, I have been talking about two able-bodied, reasonably competent people of whom mutuality could logically be expected. But there are many situations that call for caring as caregiving, and some of these situations are of long duration. They require self-sacrifice. When a husband is ill or disabled, when a child is severely handicapped, when a family is wracked by a variety of problems, the female caregiver may be overwhelmed. Situations of this sort increase our interest in a care-driven concept of justice. We should work to advance policies that give these perpetual caregivers the physical, emotional, and economic support they need. In this, care ethics is demonstrably superior to some religious ethics that glorify suffering and "help" caregivers by telling them how saintly they are. From the perspective of care ethics, suffering is to be eliminated or relieved, not glorified or ignored. A just-caring society would provide the conditions under which caring has a chance to flourish.

Care ethics views autonomy as a state of limited, appropriate, and at least minimally satisfying control. It is anchored in a relational ontology, and the self it describes is a relational self. I have described this self as a growing collection of encounters, actual responses, memories, reflections, evaluations, and acquired responses. One might think of it as an enormous set of ordered pairs (A, B) and (C, A) representing encounters of the agent A with other persons, nonhuman living things, objects, ideas, and A's own reflections. Each recorded encounter has had some meaning—some impact on consciousness—for the physical being, A. Everyday jostling in a crowd, dropped spoons, dust specks brushed off, are probably not recorded, although—who knows for sure—daily jostling and other regularly experienced situations that have no individual meaning may, as a mass, make a permanent mark. The vital point is that each recorded encounter was entered because it generated, or could be characterized by, some affect. Caring encounters are recorded *as* caring—offering or receiving care. Indeed, as discussed earlier, it is this important subset—the set of all caring encounters—that furnishes the base for A's ethical ideal of caring.

"A" is a gradually changing, aging physical entity and, as a self, A is continually changing, developing. A_i, as a representation, properly appears in a chain of encounters as $A_1, A_2, \ldots A_n$. Each new A is a product of previous As as well as the conditions of the new situation. If we could lay out, make visible, this enormous set of encounters, we would find in a caring, competent "self" many encounters with the self's own consciousness. There would be many episodes of reflection and evaluation, many encounters imagined for the purpose of assessing likely consequences.

Not all of the entries are caring. Some are relations of dislike, fear, disgust, and other affects common to human life. In some lives, there are few entries (C, A) that signify A's being cared for. If we could examine all of a subject's childhood encounters, we could predict quite accurately the direction of the subject's life and the problems that would arise. Physical violence and the emotional violence we call *shame* figure prominently in the lives of criminals. James Gilligan, a psychiatrist who has worked for years with violent criminals, claims that violent behavior is often a result of repeated blows to the young psyche; especially powerful is the deliberate and cruel use of shame as a means to control the young. Selves largely devoid of caring encounters often become violent, Gilligan warns, in an attempt to convert injustice to justice. It seems only right to these people that others should suffer to compensate for what they themselves have already suffered.[17]

We will return to Gilligan's observations in chapter 8, because beyond the provocation of violence in shamed individuals, a culture of masculinity is one that promotes violence as a way of solving problems. Gilligan comments: "If humanity is to evolve beyond the propensity toward violence that now threatens our very survival as a species, then it can only do so by recognizing the extent to which the patriarchal code of honor and shame generates and obligates male violence. If we wish to bring this violence under control, we need to begin by reconstituting what we mean by both masculinity and femininity."[18] My mathematical depiction of the relational self as a huge set of ordered pairs representing encounters is itself a mere device, and it is useful to note how the omissions and inclusions describe the actions and undergoings of the self under consideration, but as a device it has obvious limitations. To achieve fuller understanding we need to know more about the situations in which the entries were recorded.

We also need to consider the possibility that an encounter may be wrongly recorded; that is, A may evaluate an encounter as caring when an impartial observer would label it as abusive. A dramatic example of this sort of misevaluation is illustrated in the character of Ernest Pontifax (mentioned earlier in a brief discussion of pathologies of care) in Samuel Butler's *The Way of All Flesh*.[19] Repeatedly shamed and whipped (for his own good, his father said), Ernest believed for a long time that his parents really cared for him and that he was himself a miserable sinner of a son who deserved the treatment designed to save him from his sinfulness. If it were not for a wise and kindly outsider (the book's narrator), Ernest might never have developed a genuine self—one that records its own encounters with the affects actually felt.

Similarly, Louise and Tom, children of Thomas Gradgrind in Charles Dickens's *Hard Times*, were forbidden to wonder; they were instructed to acquire and use only facts. How is a genuine self to develop if the inevitable affects of life are ignored? How is empathy to develop if one is forbidden to wonder what others are thinking and feeling? And how can one learn to reflect and reevaluate if one cannot wonder?

The self we are seeking here—autonomous within a limited span—must be able to think, reflect, wonder, plan, reassess, feel, and see things with some clarity. In relations with others, the self should be reasonably competent in achieving empathic accuracy and comfortable in feeling and expressing sympathy. Such a competent self must be capable of both imagination and critical thinking, for questioning our own socialization is the main path to the limited autonomy we can hope to exercise.[20]

CRITICAL THINKING

I can make no attempt here to present a comprehensive discussion of critical thinking, a huge topic.[21] However, we must connect critical thinking with autonomy and with the ethics of care. Perhaps critical thinking is not, strictly speaking, necessary for those who engage regularly and easily with natural caring. Caring is based more directly in affect than reasoning. However, critical thinking can be useful in analyzing what to do by way of a caring response and how to do it. But this process is better labeled "means-ends" analysis rather than critical thinking, if we mean by *critical thinking* careful analysis and challenge to our socialization and how it affects our autonomy.

Critical thinking is essential in developing autonomy, but more generally it is also essential in exercising ethical caring. When something goes wrong in the usual relations of natural caring or when we move into a domain so large that natural caring is not possible, we must draw on ethical caring. Ethical caring has an urgent need for critical thinking. It finds little guidance from broad, abstract principles, and as we'll see in the next chapter, it rejects the Confucian move to an elaborate set of specific rules. It also refuses to put great reliance on the virtues of moral agents, because such reliance directs attention more to the one-caring than to the cared-for. Ethical caring requires the analysis of the situation, persons involved, needs, values, and resources available. Possibly no other moral approach has greater need for critical thinking, and it must be exercised in every facet of life. It gains even more importance applied to women's lives.

Critical thinking under this definition is part of a serious program to achieve self-understanding and to extend the span of control over our lives. It is especially important in helping us to maintain caring as a preferred moral approach to life and to spread the influence of care ethics. As a respected approach to ethics, caring is a newcomer, still developing, still uncovering both difficulties and promising applications. Because it is associated with women, and I have claimed that it is rooted in maternal instinct, many men refuse even to discuss it, and some feminists scorn it, fearing that its acceptance will aggravate the already subordinate condition of women. Earlier in this book I tried to show that much of this fear arises because of the confusion over caring and caregiving, but although the two concepts are distinct, they are clearly related, and we have to show how an ethic of care can be used both to enhance caregiving and to promote women's autonomy.

Caring and caregiving are related in the most obvious way. If someone addresses us and expresses a need for bodily care, we must provide that care, and whether we provide it ourselves or hire someone else to do it, someone will necessarily become a caregiver. Why is such work held in contempt and poorly paid?[22] The main answer to this is that the work has long been done by women, and women have been thought inferior to men. Some argue that it is the work itself that is considered inferior—involving contact with bodies, bodily fluids, excrement, and soiled linens. However, sanitation workers also handle smelly, disgusting material, and although their work is not envied or admired, they are often paid better than those who tend to the bodies of incapacitated elders or care for young children.

And what is so objectionable about working with young children? The answer to this seems simply to be that it has been "women's" work and therefore automatically devalued. Sonya Michel cites a 1989 study finding that the most important factor in quality child care is staff wages, but "the study also found that staff salaries in child care centers were 'abysmally low' compared to salaries of employees with similar levels of education and training in other occupations."[23] Some feminists believe that the way out of this continuing exploitation is for women to refuse to do such work, especially as unpaid home-workers. In particular, they want to encourage women to take paid jobs and to diminish the gender division of labor at home. To accomplish this, women must break into occupations dominated by men, but to do this requires expanded child care at affordable prices. Who will do the child care? The answer "government subsidies" begs the question. *Persons* must be paid, and the success of some women comes at the expense and continued exploitation of others unless, as recommended earlier, we insist on an adequate wage for childcare workers.

Critical thinking on issues related to women and work is essential if we are to further care ethics *and* women's autonomy. Riane Eisler is right, I think, to call for a revaluation of care-work.[24] Childcare, housekeeping, and all of the tasks required to maintain families should be neither scorned nor romanticized. They should be recognized as valuable and rewarded accordingly. In a recent book, Linda Hirshman expresses her disdain for childcare and housekeeping: "Although childrearing, unlike housework, is important and can be difficult, it does not take well-developed political skills to rule over creatures smaller than you are, weaker than you are, and completely dependent upon you for survival or thriving. Certainly, it's not using your reason to do repetitive, physical tasks, whether it's cleaning or driving the car pool."[25]

Hirshman neglects to notice that much of the paid work women do these days outside the home is repetitive, tedious, and ill paid—often done in a claustrophobic and unattractive cubicle. She is wrong again in regarding housework as unimportant and demonstrably wrong in saying that work with children does not require "well-developed political skills." Such work requires empathic skills of the highest order, and it also encourages the caregiver to continue reading and learning. The best caregivers not only meet the physical needs of their charges but also attend to their charges' intellectual needs.

I am not arguing that girls should be encouraged to abandon professional careers. Quite the contrary. Any girl (or boy) who has the interest and talent to pursue a professional career should be encouraged and supported in doing so. I am arguing that we should cultivate our capacity to care so that, in trying to eliminate gender divisions in work, we do not exacerbate class differences—a nasty phenomenon that is growing.

Utopian thinkers have sometimes described societies in which the least attractive public work is shared among all the citizens, but they seem to have little grasp of what it takes to make a home comfortable, efficient, and attractive. In Edward Bellamy's utopian novel *Looking Backward*, the narrator asks his hosts, Dr. and Mrs. Leete, who does the housework: "'There is none to do,' said Mrs. Leete. . . 'Our washing is all done at public laundries at excessively cheap rates, and our cooking at public kitchens. The making and repairing of all we wear are done outside in public shops. . . . We have no use for domestic servants.' "[26] At the time of Bellamy's writing, not having to cook or do laundry would have been a tremendous relief. However, Bellamy apparently knew little about homemaking. What about bedmaking, collecting and sorting the laundry, dusting, sweeping, carpet cleaning, picking up, shopping (if only for snacks and drinks), cleaning bathrooms, watering and feeding houseplants, washing windows, and many other small tasks? And what about all the odds and ends associated with childcare? *Looking Backward* was highly influential (and much admired by John Dewey), and its speculations on social justice are worth considering even today, but Bellamy had no real notion of what it takes to keep a house. Were he alive today, he would see that self-service laundries and fast-food eateries have not eliminated housework.

Critical thinkers must look at all sides of women's occupational problems and aspirations. There is no need to denigrate women's traditional work, and much of it still has to be done. And it is foolish to romanticize public, paid work. Every girl, with the help of careful guidance, must bring her critical intelligence to bear on decisions that will affect her occupational life and that of her sisters. In striving for an optimal level of autonomy, girls must also use critical thinking to examine their lives as students. Recent evidence shows that on average girls are doing much better than boys in school. They are taking more Advanced Placement courses, getting better grades, and achieving more bachelor's degrees.

Some social scientists argue that girls are better than boys at "studenting" because they are more self-disciplined.[27] This claim raises several interesting questions, among them whether girls today are more self-disciplined than earlier generations and, if so, why this might be the case. It is more likely that girls see increased opportunities for careers and are willing to discipline themselves at the tasks that open the doors to these opportunities—just as they were once "self-disciplined" in accepting their exclusion from these opportunities. But the interesting question in connection with autonomy is, To what extent is self-discipline autonomous? The answer to this is not obvious. Self-discipline of the sort described by Freud with respect to the superego is clearly heteronymous. The stern, moral father has been internalized. Girls have always been more compliant and obedient than boys, and it is possible that they are doing better in school today because they have been told that they can and should do better. The question is how much of this behavior can be traced to critical self-analysis and how much to compliance with authority.

A related matter of keen concern to care theorists is that many bright girls are advised to scorn the caring professions and prepare for more lucrative and prestigious careers. If girls really want careers in professions formerly closed to them, they should certainly be encouraged to pursue them. But if they are rejecting work to which they are emotionally drawn because respected authorities tell them they "can do better," then the decisions are not autonomous, and they may not be satisfying in the long run. Care theorists applaud the increased opportunities now available to women, but we do not scorn the work that our predecessors have done for centuries.

In today's schools, teachers often tell students, "Always do your best in everything," and the school then forces all of them into a narrow, prespecified range of courses. It is no wonder that some kids—often creative, independent-minded ones—just give up. The heteronymously savvy and compliant succeed. It would be far better to tell kids, "Do your best in *something*," and then invite them to explore a wide range of possibilities. Part of that exploration would involve critical self-analysis.

As women try to extend the range of control over their own lives, they might seriously consider abandoning affiliation with institutional religion. Until the great religions explicitly reject much that appears in their sacred texts, critically intelligent women should simply walk away from these institutions. All three great monotheisms have the myth of female inferiority built into them. It is not enough that they stop talking about the "rightful" or "natural" domination of women by men; they must explicitly and apologetically reject all such doctrines and stories. There is little point in encouraging critical intelligence if we are unwilling to exercise it in domains that continue to keep women subordinate.

In this chapter, we've looked at autonomy from several angles. There may indeed be no autonomy as philosophers and theologians have defined it. The best we can hope for is intelligent heteronomy and some limited control of our own lives achieved through critical intelligence. I tried to supply reasons why care theorists should encourage critical thinking in girls—especially critical self-analysis. However, I also noted that it is possible to live a life of natural caring without exercising a high level of critical competence, because caring is inspired more by feeling than by reason.

We turn next to a deeper discussion of relation, virtue, and religion.

FIVE

Relation, Virtue, and Religion

Care ethics shows several important similarities to other approaches to moral theory, and some of these similarities have received considerable attention in the past few years. We will look first at the similarities and differences between care ethics and virtue ethics, then at those between care and Confucianism, and finally at those between care and Christian agape. The chapter will conclude with a critique of women's involvement in religion.

VIRTUE ETHICS AND CARE ETHICS

Virtue ethics and care ethics share a significant basic characteristic. Both turn to something inside the moral agent instead of to a principle when faced with making a moral decision. The virtue ethicist depends on the character of the moral agent; the care theorist turns to the ethical ideal established over a lifetime of giving and receiving care. Both are concerned with moral relations—with how we should meet and treat one another. But care theory puts greater emphasis on the relation, virtue theory on the moral agent.

From an evolutionary perspective, relation is clearly prior to virtue. It is in the mother-infant relation that the virtues associated with motherhood are developed. When a female gives birth, she is defined—for a time at least—as a mother, as a being related to another by an intimate connection. And the child is dependent for life itself on a mother—its own or one who plays that role. As we move beyond the original condition and its basic relation, our very selves are formed in relation. As Martin Buber put it, "In the beginning is the relation."[1]

Contemporary feminism puts great emphasis on relation. Virginia Held, for example, writes: "The ethics of care attends especially to relations between persons, evaluating such relations and valuing relations of care. It does not assume that relations relevant for morality have been entered into voluntarily by free and equal individuals, as do dominant moral theories. It appreciates as well the values of care between persons of unequal power in unchosen relations such as those between parents and children and between members of social groups of various kinds. To the ethics of care, our embeddedness in familial, social, and historical contexts is basic."[2] Other feminist philosophers also argue that the conception of persons as relational selves is fundamental.[3] But communitarian philosophers, too, put emphasis on relations and how selves are formed by the practices and beliefs of family, community, and nation.[4] How do their views differ from those of care theorists? A thorough investigation of this question is beyond the scope of the present project, and it is possible that the ethic of care, as I am describing it, could be developed as a form of feminist communitarianism. However, given the history of communitarianism with its exclusive emphasis on male experience and its neglect of the most basic relations in favor of societal practices, I would be reluctant to give care ethics such a name. Readers should keep in mind that the eventual goal is convergence, a blending of views that will contribute to the well-being of all human beings, not a separate community life for males and females. Care ethics is by no means restricted to women—any more than Kantian ethics or Utilitarianism should be restricted to men—but it grows out of female experience. It is an approach to moral life that has only recently been recognized in articulate form, and time is required for its full articulation. In any case, my sense is that care ethics will not (should not) embrace the sort of traditional authority claimed by communitarianism.

In a perspective that claims relation as basic, the emphasis is necessarily on dyads, and this emphasis generates another—on reciprocity. Buber writes on reciprocity: "Relation is reciprocity. My You acts on me as I act on it. Our students teach us, our works forms us. . . . How are we educated by children, by animals! Inscrutably involved, we live in the currents of universal reciprocity."[5] Buber's reciprocity is not the contractual reciprocity familiar to us in liberal philosophy. As I have interpreted it, reciprocity is recognition, a positive response to a carer's efforts to care. It is this response of the cared-for that completes a caring relation or encounter. The response need not be one of gratitude. Indeed, in many situations, gratitude would be inappropriate. The response is one of recognition or acknowledgment, and the caring relation is itself one of mutual response.

The infant contributes to the parent-child relation by smiling, cooing, wriggling, and extending his arms. Students respond to their teacher by listening, asking questions, pursuing projects with increased vigor. Patients contribute to the nurse-patient relation by showing their relief from pain. Traditional moral philosophy has either ignored these contributions to moral life or interpreted them as virtues of some sort. A relational ethic is not so bound. It is interested in moral *life*, not only in moral reasoning, moral principles, and moral virtues.

Critics could argue, of course, that the responses just mentioned are heartily welcome but that they have nothing to do with morality unless they are deliberately chosen in response to duty, principle, or some virtue of character. This criticism is a predictable consequence of the individualistic thinking that has long characterized traditional and abstract moral theory. It too is deeply interested in relations; that is, it is interested in how we should meet and treat one another. But it does not make relation basic. It looks analytically at individuals, at how they construct relations, make choices, and conduct themselves in relations. It rarely recognizes relation as the incubator of both individuals and virtues.

In contrast, an ethic of care begins with relation. It is not so much interested in the moral credit due to each individual as it is in the strength of caring relations and health of the entire web of care. It is obvious that responsive infants, students, and patients contribute to caring relations. If we insist on a *moral* contribution, we can claim it in the support such responses give to the continuing efforts of carers. Carers draw energy from the responses that complete a caring encounter. Their empathic capacity may be enlarged: the power to read others accurately is confirmed; their choice of action in motivational displacement is recognized; their sympathy is augmented by a sympathetic harmony in the cared-for. But this sort of moral contribution is not usually consciously chosen by the cared-for as a moral agent. The infant does not choose to make a moral contribution. The student is often unaware that his increased interest has contributed to the teacher's continued commitment to care. And the patient's sigh of relief merely signifies a reduction in pain.

Positive responses in at least some of those cared-for also tend to support carers when other cared-fors do not respond. Teachers working with difficult students may be sustained in their efforts by the contribution of those who do respond. Indeed, it is the realization that one's efforts often produce caring relations that inspires a carer to draw on ethical caring when natural caring fails. Consider what would happen (and sometimes does happen) when no one responds to a carer's attempts to care. We call the result burnout. Or in the language we are using here, we might speak of empathic exhaustion. The carer gives up and may even say, "I don't care anymore."

Colleagues and friends also help carers to draw on their ethical ideal of caring when natural caring fails. Michael Slote discusses two ways in which a second person (one outside the relation in question) can encourage a carer to call upon her ethical ideal.[6] One way is to express feelings resembling those usually experienced by the carer, thus inviting her empathic reaction. A second is to remind the carer of how she usually responds in natural caring. As Slote points out, the two approaches often interact, and both may be enacted in a series of conversations. When, for whatever reason, we do not get the sustaining response from those we try to care for, we need to be propped up by others around us, who remind us of our best selves. This is a powerful example of confirmation. Those who know us well enough to attribute the best possible motive to our actions help us to rethink and to choose means more compatible with the ends we seek. Helpful others remind us of our better selves.

There is another reason for giving more attention to the role and influence of the cared-for. Caring, as it is described in care ethics, is not universal (that is, it is not exhibited by every individual everywhere), but the desire to be cared for *is*. Some deny this by declaring that they don't want to be "cared-for"; they say, for example, that they just want respect. But, of course, the need for respect—like every other expressed need of cared-fors—is detected by a skillful carer. It is one way of wanting to be cared for. *Caring*, as described by care theory, is not merely a fuzzy feeling, nor is it a prescription for how all cared-fors must be treated. It is a moral response to expressed needs. Every human being wants to be recognized in some way—to be protected from harm, to be seen as fully human, to be respected, to be comforted, to be fed.

People who are prepared to care respond empathically as described earlier: they listen, read others effectively, feel sympathy, and experience motivational displacement. They are moved to reevaluate their own readings and to see lovingly, and they are motivated to draw upon their ethical ideal of caring. Now, what of caring as a virtue? Relational carers acknowledge the use of "caring" as a virtue. Someone who regularly establishes or maintains caring relations might be described as a caring person. The caring relation is primary and provides the necessary foundation for the attribution of caring as a virtue.

There is another way of looking at caring, however. Some well-intentioned people might be labeled *virtue-carers*. These people often decide a priori what others need, and they respond conscientiously to these inferred or assumed needs. Many parents, teachers, and other authority figures fall into this pattern. At the extreme, virtue-carers may be self-righteous and dictatorial. Most of us have known teachers like this, and sometimes we have loved and admired them. They seem to know what is best for their charges, and they work hard in what they see as the best interest of those for whom they care. This way of "caring" contrasts sharply with relational caring. However, it is fair to say that most of us behave this way at times. Relational caring and virtue-caring name dominant—not exclusive—approaches to caring.

In the virtue-caring approach, carers see caring as a virtue, that is, as a morally admirable trait of character. Virtue-caring teachers are likely to insist that things be done as they say and on time, and they are less likely than relational carers to attend to students' expressed interests. We have to be careful here, because very few people are totally consistent in behaving as either relational-carers or virtue-carers, but the distinction is useful and holds up across broad lines. In teaching, relational-carers are likely to adopt a theory of motivation that assumes students are already motivated, and the job of teachers is to discover what this motivation is and use it to move students toward intellectual growth. In contrast, virtue-caring teachers may hold a theory of motivation that requires teachers to motivate students to learn the prespecified material. This difference has been identified as one between progressive and traditional educational philosophies, but it may also be characterized more generally as one between relational-caring and virtue-caring. Readers should note that, for the most part, schooling at every level is more nearly defined by virtue-caring than by relational-caring. Curricula, textbooks, and syllabi are all constructed before teachers have actual interaction with their students. We might all agree that well-educated, competent adults should have some idea of what subjects students should learn and what is most valuable in each subject, but relational-carers use the prescribed materials more as a guide than a script. They will offer choices and options and encourage students to take charge of their own learning.[7]

Similarly, parents who embrace relational-caring will regularly engage in cooperative decision making with their children. Suppose Ms. A's son, Bob, protests that he does not want to take algebra in high school. He wants to be an artist, let's say. Ms. A listens. She may respond in any of several ways. As his mother, Ms. A knows Bob well. She may be convinced that he really will find a satisfying future as an artist. She will talk to him at length about the possible consequences of his choice but allow him to opt out of algebra. Or, knowing that Bob has talents that might lead to architecture or design, she may coax him to give algebra a try. Assuring him of help and understanding, she may get Bob to agree that it is in his best interest to sign up for algebra. Ms. A will not, however, insist at the outset that she knows what is best for him and that he *must* take algebra—end of argument.

American schools today act almost entirely in the virtue-caring mode. Many schools—even some entire states—now insist that all students, regardless of interests or talents, take college preparatory courses. This is done in a spirit of generosity, supposedly ensuring that all students get the best possible education. But it devalues the essential work that is done by people who do not get college degrees, and it ignores the interests of many students. It also establishes a definition of education in terms of mastery of specific subject matter as opposed to the quality and capacities of the fully human persons it produces.

On the international level, nations and organizations often act as virtue-carers. When we express certainty about the superiority of democracy, we may feel righteous about forcing it on others. Like the virtue-caring teacher who says, "Someday you'll thank me for this," when forcing us to do things we hate, nations sometimes believe that all people cherish individual freedom and will thank us for forcing it on them. Obviously, virtue-caring is not the sort of caring endorsed by care ethics. It might better be considered a form of paternalism.

Virtue-caring, as I have described it, may not be enthusiastically endorsed by virtue ethicists either, and I do not mean to suggest that they cannot respond adequately to the problems raised here. Virtue ethicists might well modify the behavior of virtue-carers by bringing other virtues into play, and indeed many virtue-carers display virtues such as patience, compassion, and conscientiousness. However, virtue-caring and virtue ethics share a common emphasis on the character of moral agents and their exercise of virtue, and virtue ethicists have to show how the picture of virtue-caring I have drawn can be either supported or avoided.

In the last few years, there has been lively debate over whether care ethics should be subsumed under virtue ethics. Although the two ethics share certain features, they are rightly considered as basically different approaches, largely because a relation of natural caring precedes the development of virtue. A second difference is that, for care ethicists, *caring* is used most powerfully to describe relations, and we speak more often of caring relations than of caring agents. When it is used as a virtue, it points to a person who regularly establishes and maintains caring relations. It does not consider the virtue of caring as the possession of an individual moral agent. A third difference lies in the conception of relation. For care theorists, many significant human relations may simply happen through birth into a particular family or through encounter; not all relations are chosen. Bumping into a stranger is an encounter or minimal relation, and it may or may not be a caring encounter. Similarly, a parent-child relation may or may not be a caring relation. Virtue theorists, in contrast to liberal philosophers, largely agree with care theorists on this; many of our most important relations are not formed by choice. However, in the work of some virtue ethicists, there is more emphasis on roles and status than in care theory.

One reason for exploring the possibility of subsuming care ethics under virtue ethics is to avoid perceived difficulties in care ethics. Several critics have expressed fear that care ethics (as I've developed it) is susceptible to corruption.[8] Recall my 1960s story of Ms. A, who admired and sympathized with her black graduate student classmate Jim.[9] When he spoke eloquently in class against the injustice his people suffered, Ms. A was deeply moved, but when he spoke of "going to the barricades," she was unhappily aware that, if things got so bad as that, she would stand with her bigoted relatives against Jim and his people—even though she knew Jim's people were right. This story aroused indignant criticism.

However, as noted earlier, I was not prescribing what Ms. A *should* do but, rather, taking into account what she *would* do. Caught in a terrible situation—having to protect her family, even though they were clearly wrong, against strangers who were clearly right—she would choose to stand with her family. In telling the story, I reminded readers that this is the way we humans are; this is reality. When we are trying to develop an ethic, we have to pay attention to reality. I ended the story by saying, "Everything must be done, then, to prevent 'going to the barricades,' for if that occurs, diminution of the ideal is inevitable."

Commenting on this story and what is regarded by some critics as Ms. A's morally deplorable choice, Raja Halwani recommends that care ethics be subsumed under virtue ethics so that agents would be called upon in such situations to make virtuous decisions: "And this is surely a plausible demand as far as, specifically speaking, caring actions are concerned: no matter how spontaneous and emotional we want caring actions to be, we do, and should want them to issue from reflection and morally committed agents. Otherwise, we will all be in danger of becoming Ms. A's."[10] But we *are* all Ms. A's. That was exactly my point. We are biologically constituted to stand with our own; the closer the biological relation, the more closely we stand. In his notes, Halwani wrongly explains my position by writing that, for me, "caring for her relatives constitutes the supreme moral imperative. . . . But in not fighting on Jim's side, Ms. A not only neglects considerations of justice, but is also willing to violate her integrity, given that she believes that Jim is correct in his demands."[11]

I did not claim that Ms. A's choice was morally admirable. I wanted to remind readers that, morally, we must do all we can to prevent situations such as these from occurring. As in the Trolley and Lifeboat dilemmas, conditions have thrown us outside the usual domain of morality. The only truly moral conduct over which we have some control is exercised in the events and choices preceding such emergencies. When these terrible situations are not prevented, we will stand with our own. We will be forced into a situation in which neither choice will leave our moral ideal intact. Consider how citizens rally around their own nation—no matter how wrong it is—when that nation comes under attack. I'll say more about this, too, in chapter 8.

Now, clearly, as Virginia Held has advised us, we are moral subjects, and we should not be unreflective instruments of our biological nature. When we collaborate with an act that is morally wrong, we cannot use as an excuse that it is the nature of human beings to behave this way. That's why it would be a mistake to push the naturalization of care ethics too far. Ms. A's decision, however, should not be judged either morally right or morally wrong. Standing with our own—protecting them, correcting them where possible, but "staying with"—is a foundation of morality. There are many human tendencies that we can control, even overcome, but there are others that are so deeply entrenched that we have built whole mythologies to support them. Much of national patriotism is in this category. We have created songs, rituals, poetry, stories, and whole histories to justify and inspire "standing with our own." The case of Ms. A (and many similar cases) is one of human tragedy. Her moral responsibility is not to fight against her own people but to do all she can to overcome the injustices that lead to fighting.

An emphasis on the virtue of individual moral agents also leads to a distinction between altruism and egoism that care theorists find questionable. Because we are defined in relation, when we care for another, we may strengthen the relation and thereby benefit ourselves as well. A distinction between altruism and egoism can be maintained at the extremes—sacrifice of one's life, for example—but in the broad middle range of human interaction, most acts that would be labeled altruistic by virtue theorists are more naturally described as acts that maintain or enhance the caring relation.

We might also describe caring as a complex virtue; that is, a person deserving the label "caring" by reason of the relations she regularly establishes, must exercise certain desirable characteristics that might be called virtues. Primary among these would be *sympathetic attention*. I will defer further discussion of these component virtues until chapter 6.

CARING AND CONFUCIANISM

Confucianism is sometimes considered a form of care ethics. Chenyang Li, for example, points to several features shared by Confucianism and care ethics: their noncontractual nature, the deemphasis of principles, and their theoretical beginnings in family life.[12] But as Daniel Star pointed out in a critique of Li's article, these features are often found in virtue ethics, and Confucianism is rightly classified as a form of virtue ethics.[13] It is not on that account necessarily a type of care ethics. As we have already argued, care ethics is distinct from virtue ethics, although the two share some characteristics.

Confucianism, like care theory, does express a distrust of universal principles, mainly because it holds that ethical problems must be considered and resolved in the context in which they arise. It is a form of particularism. However, despite the deemphasis on principles, Confucianism has developed a large set of rules, *li* and *yi*, governing what is socially right and proper. Care theory resists this move.

It also seems right to say that the two ethics look at virtue itself somewhat differently. Both start with the family and work outward past the inner circle, and I have described the mother-child relation as an "incubator" of various virtues. However, the caring relation remains central to the fully developed ethic, and even the definition of caring as a virtue depends on the regular establishment and maintenance of caring relations. In contrast, the prominent Confucian philosopher Mencius starts with a concept very like caring (*ren*, or *Jen*) but uses it to arrive at a more fully moral *virtue*.[14] In care theory, this move would place ethical caring above natural caring, and I have argued that the reverse is true. Natural caring is to be preferred, and ethical caring is invoked to establish or restore it.

Confucianism, like communitarianism more generally, is tradition-oriented and, because all human traditions have been articulated by men, it is male-oriented. Li admits that Confucianism has been used to oppress women, but he suggests that it might widen its domain of application much as Western liberalism has done to include women and to condemn slavery. I will return to this issue in a bit. Insofar as Confucianism is often embraced as almost a religion, this claim is doubtful. As we have already noted, religions sometimes stop "talking about" questionable doctrines once fully embraced, but they rarely reject them outright and admit that they are wrong.

The fact that traditions have been formally described by men does not imply that there are no distinctly female traditions. Indeed caring for particular others may be clearly identified as a female tradition. The present task is to further its articulation as an approach to moral life. Star points to the work of anthropologist Margery Wolf to show that women in rural China, lacking the well-defined sense of self enjoyed by men through patriarchal connections, exercise what looks to be an ethic of care—responding to others in accordance with their expressed needs.[15] Like male Confucians, the women break rules when the needs of the cared-for seem to require rule-breaking. However, their rule-breaking is just that—breaking or ignoring a rule for the sake of the cared-for; it is not a prelude to rewriting or reforming the rules. Under the male domination of Confucianism, women do not have the power to do this, and the particularity of their decisions is partly a sign of some independence within the larger environment of subordination.

Confucianism is demonstrably different from care ethics in its emphasis on prescribed roles and duties; further, the central duties of Confucianism are male-defined: son to father, citizen to ruler, wife to husband. Caring, phenomenologically elaborated, does not prescribe obligations in terms of roles. As we have noted, however, it does recognize our biological nature, and it takes this nature into account in developing care ethics. Like Confucians, we recognize the centrality of family and blood relations, but we do not justify our actions entirely in terms of family loyalty, nor do we accept a role-identified system of authority. I have already noted, in the case of Ms. A and her black classmate, that Ms. A will stand with her bigoted family, but I did not attempt to justify her decision morally. Faced with a theoretical encounter at the barricades, we should recognize a turning point. It seems equally questionable morally to fight against one's own people in the name of universal justice or to fight with them against those who have legitimate objections. Care ethics locates our moral obligation in all of those acts and events that lead to or prevent confrontation at the barricades and encourage fighting. It is to these points that we should return and concentrate our moral efforts.

Philosophers, theologians, and other theorists are often unwilling to recognize moral dead ends and the need to acknowledge turning points. War is a prime example. Political philosophers continue to recommend that people find a way to establish and live by rules of just warfare. But even when the cause of war is just, war is never prosecuted justly. When we contemplate war, we face a moral dead end and must turn back or face the fact that our ethical ideal will inevitably be diminished. Our moral obligation, therefore, is to prevent war.

Let's return briefly to Confucianism's endorsement (or acceptance) of male domination. It may happen, as Li suggests, that Confucianism will gradually embrace female equality or at least not speak out authoritatively against it. But as I will argue in the last section of this chapter, if a religion (ideology or philosophical school) does not explicitly reject the doctrines that support domination, these doctrines remain quietly embedded in the religion and are thus available to those who would reactivate them. Confucianism would have to admit that it was wrong to claim the inferiority of women and, then, patiently locate and repudiate all the doctrines based on this claim. This will be extremely difficult for an ethic that puts such emphasis on social roles and tradition. Western liberalism has not yet achieved a state of full gender equality. That is a function that might have been served by the Equal Rights Amendment proposed in the United States. Instead of insisting that "all men" really includes women (but somehow people missed the intended meaning), legislators would have to admit that their forebears were wrong and that it is time to recognize and reject a vocabulary that denigrates females. We would have to admit also that the invitation to women to join men and be more like them suggests an invitation in the opposite direction. Perhaps there are ways in which it would be good for men to be more like women.

CHRISTIAN LOVE

There is a strong line of thought in Christianity that calls upon Christians to love one another and even to love their enemies. Christians have a duty to respond to pain and suffering with compassion, to poverty with charity, and to evil with forgiveness. Christian history is filled with stories of saints and ordinary people who have obeyed the call to love and to care for others. These stories are deeply moving, and one cannot shrug aside the power of religion to evoke a form of caring.

Ruth Groenhout has given us an interesting comparison of caring and Christian love as described by Augustine.[16] Like care ethicists, Augustine expressed respect for bodily love and the centrality of concern for those close to us—family, neighbors, community members. Indeed, Augustine seems to have hinted that the love of neighbors and proximate others might serve as a cradle for growth in love of God. This sounds very like the claim of care ethics that the caregiving required in maternal activity acts as an incubator of the attitudes and virtues associated with caring.

But the caring prescribed by Christianity is significantly different from that described by care ethics. As we saw in the earlier discussion of Simone Weil and attention, Christianity construes the duty to care as a commandment from God, whereas care ethics depends on an internal "I must" that arises spontaneously in response to needs expressed by individuals, and in ethical caring on a treasured store of memories of caring and being cared for. Groenhout recognizes the dramatic difference between the views of care ethics and Augustinian doctrine on authority. Augustine construed authority as domination and insisted on submission .as the required response to legitimate authority. Faced with the dictates of legitimate authority, the Augustinian subject obeys—gives up the quest for individual power and the pride that might accompany its realization. In contrast, carers try to use their power to encourage the reflective intelligence of those for whom they care. This is an important difference. Moreover, in care ethics, a moral agent may ignore the inner voice of natural caring or of the ethical ideal, but if she listens to either one, she will never hear a message endorsing violence or cruelty.

Sadly, the message coming from the Christian God is not always one of love. In the New Testament, there are many passages threatening damnation, destruction, and endless suffering. Redemption is promised, but it is reserved for those who believe that Jesus is Lord and Savior. What of those who, for whatever reason, cannot believe this?

The Christian message of love is also premised on the idea that God is a loving God. What evidence is offered for this extraordinary claim? It is a claim that angers concerned atheists and agnostics, and it has driven many believers to give up their faith.[17] The God of all three great monotheisms has a history of rage and cruelty. On the one hand, he exhorts his followers to treat others with compassion and charity, but on the other, he often encourages them to exclude or even destroy those who reject the beliefs of the group through which he speaks.

The mixed messages from God are a product of many inter-preters. We do not encounter God-in-person as we do our human mothers. In a book by Yossi Halevi exploring the pos-sibility of reconciliation among the three monotheisms, we hear these contradictory messages over and over again. We hear also from the faithful all sorts of beliefs that cannot be substantiated save through personal witness: "'God has the last word.' 'Some-times God gives you refusals.' 'But I believe that God has some-thing special in mind for this land. It's God's secret.' 'God doesn't repeat Himself.'"[18] The end result of reading this book is a combination of great admiration for the deeply religious people who seek reconciliation among irreconcilable beliefs and a nearly hopeless despair for the future of the Middle East. What people "hear" from their God often suppresses the voice of those who would be cared-for, and expressed needs may be righteously ignored.

This ambivalent discussion of Christian love might be dis-turbing for some readers, and so it may be helpful to conclude this section with an example of the differences I've identified. Simon Wiesenthal has told a story—a fictional autobiography of sorts—in which a young Jewish prisoner in a World War II concentration camp is taken to the bedside of a dying, blinded Nazi soldier.[19] The young soldier, Karl (just twenty-one), expressed the need to confess a terrible crime before he died. A year earlier, he had been part of a company of soldiers ordered to burn a structure filled with Jewish villagers and to shoot any who tried to escape. Karl had obeyed orders. Now, on his death-bed, he sought the forgiveness of a Jew. The young Jew (osten-sibly Wiesenthal) listened and finally, without a word, rose and left the room.

Thirty years or so after the incident just recounted, Wiesenthal convened a symposium and asked participants to comment on whether he had been right or wrong to walk out on Karl without comment, without forgiveness. Understanding what Wiesenthal had suffered, none of the symposium participants would pass judgment on his decision. Nor will I. But the participants expressed themselves on what, ideally, should have been done or on what they would wish they would have done in such a situation. The responses dramatically illustrate the differences I've been discussing.

Several argued that forgiveness should have been granted. Some argued for this conclusion on the basis of religious principle, and some used religious exemplars—saints—to illustrate the favored response. The principles and saints mentioned would not have been unknown to Karl. He had a Catholic upbringing, but it did not immunize him against the infectious power of the Nazi education he experienced in school and in the Hitler Youth. Karl's case shows vividly the danger in Augustine's endorsement of hierarchy and authority. When a person is instructed to submit unquestioningly to authority, the good learned from one authority may be lost when another authority—mistaken or evil—takes over. A similar warping happened in the lives of many who accepted Kant's dictates on duty but mistakenly detached them from the overriding principle of reason on which Kant built them.

Others argued against forgiveness. Some said that Simon could not properly act as a representative of all Jewry; he was called upon *as a Jew* to listen to Karl's confession and somehow recognize his repentance. How easy it is to move beyond the situation *right before us* and turn to generalities and high-flown rhetoric to escape the horrific responsibility to respond face-to-face. We can agree that Simon could not forgive in the name of all Jews. But he was addressed also as a person. Only Simon and Karl were present. A mother-carer sees two young men—little more than boys—both suffering. One suffers from the unjust acts committed against him; the other suffers from the unspeakable act he himself committed. The loss of life and freedom is heartrending, but so is the loss of moral identity. The one-caring, in near despair, would comfort both suffering boys.

In a similar line, some participants claimed that Simon had no authority to forgive. He was not God, not a rabbi or a priest. He was not himself one of Karl's victims. He was just a captive present at the scene of a repentant criminal's suffering. The one-caring says that is enough. Respond to the suffering.

Some respondents, refusing forgiveness, worried that their forgiveness might encourage similar horrors in the future. If Karl were forgiven, this might lead others to believe that they could commit the vilest of crimes and yet be somehow reabsorbed into the circle of the saved. This line of reasoning illustrates again the traditional temptation to reach beyond the real situation to generalities. If Simon had responded to Karl's expressed need for forgiveness, no one else need ever have heard about it. When no one else in the web of care will be hurt, the one-caring responds to the immediate need.

One participant, David Daiches, took a position close to that of care ethics. He put aside formulaic efforts at forgiveness and suggested instead a response of understanding and comfort, even as he acknowledged the remaining moral problems as deeply troubling.

Called upon to care, we would respond with some form of sympathy and comfort to Karl. Perhaps we would simply hold his hand. Perhaps we would cry with him—shedding tears he could not see. Perhaps we would talk about the horrors of corrupt education and obedience to authority. Perhaps we would share tales of happier times. Perhaps we would say nothing but, through touch, provide a bit of comfort. We would not ignore the face-to-face expression of need.

Christian love is often expressed beautifully, and many human beings have been motivated to live charitably by its teachings. But like other traditional approaches to morality, it reaches "beyond" for its source and justification, and the source is inconsistent. Care ethics looks directly at the human other who addresses us. The ethic of care differs from Christian caring, although they may sometimes culminate in the same activities of caregiving. In embracing care ethics, we cannot claim to have saved humanity from moral error. There is no triumphalism here. But we can claim to be in touch with the real world and an observable human nature. That should be an improvement over mixed messages and imaginary voices.

RELIGION AND WOMEN

So far, care ethics has been distinguished from virtue ethics, Confucianism, and Christian love. In care ethics, there is no need for God as intermediary, model, or enforcer. However, because religion has had such enormous influence on women's lives, we must say more about the possibility of freeing women from its domination.

We might ask, more generally, why religion has developed universally. If it is a product of evolution, what is its survival value? Perhaps, as Richard Dawkins has suggested, religion has arisen as an evolutionary by-product; that is, religion may be a by-product of a characteristic that does have survival value.[20] Dawkins suggests as a possibility the survival value of childhood trust and obedience. In order to live and eventually reproduce, children have had to listen to adults, believe their warnings, and obey. The tendency to seek and maintain the authority of religious figures might well be an offshoot of this childhood mentality. As already noted, however, reliance on an authority that initially provides protection and the assurance of love can easily be perverted to a reliance that allows one to escape personal responsibility. Acknowledging that we do not yet have an answer to this, Dawkins invites readers to explore other possibilities.

Paul Bloom discusses the inherent tendency of humans toward dualism. Apparently, most of us—at least some of the time—feel an inner spirit in ourselves, something different from our physical bodies.[21] We see the world made up of bodies and souls. Bloom uses the results of studies of child development to explain how it is that so many of us remain lifelong dualists. Perhaps, because we have thoughts not always shared with others and become aware that others also have hidden thoughts (minds), we identify these thoughts with spirit or soul—a part of us separate from our physical bodies. These thoughts are the source of our intentions, and we try to discover the intentions of others. Children tend to see an intention or purpose behind everything, asking repeatedly, What is it for? This tendency, too, might extend to adult thinking about God and natural phenomena.

Why did God cause the hurricane? Why did God save our family?

David Linden, in his discussion of the "accidental brain," explores the possibility that the human propensity for "creating coherent, gap-free stories . . . is part of what predisposes humans to religious thought."[22] One can see how the tendencies to seek intention in everything, to confuse mind with soul, and to create stories to explain it all might well lead to religious thought.

Notice that nothing discussed in this section is intended to affirm or reject the existence of God or, more generally, spiritual beings. My target is religion as it has developed in cultures invented and described by men. If humans are, by evolutionary development, inclined to believe in souls and spirits, to see intention in everything, and to tell stories to explain natural events, that still doesn't explain why religions—especially the three great monotheisms—have taken the form with which we are familiar, and it certainly does not explain why females would accept—sometimes eagerly—religions that explicitly describe them as inferior.

For an explanation of this female compliance, we turn back to the second evolutionary source of women's empathic capacities. When females accepted the protection of males, they also accepted their own subordination. Then cultural evolution, religion, and socialization reinforced that acceptance—in a sense, made it holy. But why do women today not rebel against religious institutions? Perhaps, again, submission is an accepted way to escape personal responsibility.

It seems unlikely that either women or men in large numbers actually believe the doctrines propounded by their religious denominations. As many thoughtful writers have pointed out, study of other religions and other denominations within a religion is likely to induce reflection on one's own beliefs. How can other people believe so many silly things? But when we turn to look at our own beliefs as outsiders might, we begin to suspect that ours too might be silly.

We do not have to heap scorn on this silliness, as so many outspoken atheists do today, but we have to find a way to talk about our beliefs and examine them rationally.[23] Martin Gardner, a philosophical theist, finds it remarkable that Protestants in mainline denominations still recite the Apostle's Creed. Visiting a Methodist church, he listened with astonishment to the recitation: "I would have staked a sizable bet that more than 80 percent of those present considered the creed sheer nonsense."[24] Gardner is not making fun of those who believe in God; he is himself a believer. He is pointing with chagrin at the unnecessary baggage of nonsense carried by so many believers. My own guess is that church members do not think the creed nonsense; they simply do not think about it at all. It is just part of a ritual.

In a similar vein, Daniel Dennett notes that people seem to "believe in belief"; that is, that it is somehow, perhaps morally, desirable for people to believe in God, and that we can be confident that someone believes if he or she belongs to a religious tradition. Belief is assured by membership in some religious institution. However, Dennett remarks, "Belief in belief in God makes people reluctant to acknowledge the obvious: that much of the traditional lore about God is no more worthy of *belief* than the lore about Santa Claus or Wonder Woman."[25] My concern is that much of the nonsense promulgated by religious institutions encourages the continued exploitation and denigration of women.

Consider the story of Adam and Eve and the Fall. Paul Ricoeur comments: "The harm that has been done to souls, during the centuries of Christianity, first by the literal interpretation of the story of Adam, and then by the confusion of this myth, treated as history, with later speculations, principally Augustinian, about original sin, will never be adequately told."[26] Women and men have both suffered harm from this abysmal tradition, but the harm to women has been far greater. Mary Daly, in an angry passage, notes that the male tradition has attempted to steal everything from women: "That [traditional] language for millennia has affirmed the fact that Eve was born from Adam, the first among history's unmarried pregnant males who courageously chose childbirth under sedation rather than abortion, consequently obtaining a child-bride."[27]

Daly goes on to describe the rituals established to further the subordinate role of women, and she shows how religion has upheld a sexual caste system that subordinates women: "Sexual caste is hidden by *ideologies* that bestow false identities upon women and men. Patriarchal religion has served to perpetuate all of these dynamics of delusion, naming them 'natural' and bestowing its supernatural blessings upon them. The system has been advertised as 'according to the divine plan.'"[28] Daly spends considerable time in demonstrating that the myth of Eve and her fault in bringing evil into the world is still active in both religion and literature. Some mainline believers respond to this criticism by pointing out that their churches no longer engage in such talk. They no longer treat the story of Adam and Eve as history or blame women (through Eve) for the existence of evil in the world. But no longer talking about these things is not the same as repudiating them. What is required from church authorities is a frank confession that their predecessors were wrong and that the church is now pledged to truth and equality. An apology is required. This has not happened and, given the massive volume of material containing masculine language and dedicated to the worship of a male deity, such "liberal" transformation is probably hopeless. J. Anthony Phillips comments: "This [liberal] project must eventually come to terms with the fact that the language of Bible and Church doctrine is so closely bound up with the presupposition of a male deity that all pretension to reconstructing biblical or traditional theology must be given up. Liberals will discover, through the story of Eve, that feminism cannot effect reform, only revolution."[29] It seems to me that Phillips is right on this, but the problem extends well beyond official church doctrine. Although feminist theologians have been writing for more than three decades, their work is rarely mentioned by male theologians, critics, and historians. Even when condemning religion, male writers ignore the powerful criticism of female thinkers. The only possible solution is to entirely reject the whole tradition built on the notion of a male deity. The "revolution" must not be violent. It should require turning away from corruptible authority and toward a new life of responsible commitment.

It should not be so difficult for thoughtful women capable of critical thinking to decisively reject a tradition that has so wounded them. As Richard Dawkins has pointed out: "The God of the Old Testament is arguably the most unpleasant character in all fiction: jealous and proud of it; a petty, unjust, unforgiving control-freak; a vindictive, bloodthirsty ethnic cleanser; a misogynistic, homophobic, racist, infanticidal, geno-cidal, filicidal, pestilential, megalomaniacal, sadomasochistic, capriciously malevolent bully."[30] We need not shout this language at believers, but we can use bits of it in appropriate situations to explain our own position—a willingness to explore spiritual possibilities, to listen and sympathize with traditional believers, but to stand firm in rejecting the male god.

There are thousands of books for and against religion, on the history of religion, on the differences among religions, and on the connections between religion, politics, education, and everything else. My major concern here is twofold: to make clear that care ethics does not depend on religion and to expose ways in which religion has actually impeded the development and articulation of an ethic of care.

We have already discussed the common confusion between caring as an approach to moral life and caregiving as an occupation. Religion has exacerbated this confusion by speaking regularly of caring as the provision of care—as caregiving. Out of this confusion comes an idea of caring directed by God and of caregivers as servants of God. Concentrating on what is provided by caregivers to recipients of care, there is rarely mention of receptive attention to the living other (attention is, rather, under the direction of God), sympathy for what *individuals* (not only the "poor," "sick," etc.) are going through, the need for practice in developing empathy, and the centrality of relatedness in human life.

The idea of serving also contributes to the continuing subordination of women. From the religious perspective, it is a God-given attribute or duty for females to provide care, and again this means caregiving. The view of females as caregivers is an essentialist position, whereas the view expressed by care ethics is evolutionary—subject to modification. In the religious perspective, there is no recognition of natural and ethical caring as a general and powerful approach to moral life. When caring is mentioned in connection with moral life, it is sometimes labeled a feminist ethic, but when feminists consider it, they often reject it because feminism is supposed to advance the condition of women in the public world and *caring* is likely to retard that advance. Religion, in its sacrificial emphasis on providing care, sustains this misconception.

Virginia Woolf described the dilemma well in poking some fun at the myth of the Angel in the House:

> It was she who used to come between me and my paper when I was writing reviews. It was she who bothered me and wasted my time and so tormented me that at last I killed her. [The Angel in the House] was intensely sympathetic. She was immensely charming. She was utterly unselfish. She excelled in the difficult arts of family life. She sacrificed daily. . . . [She] was so constituted that she never had a mind or wish of her own, but preferred to sympathize always with the minds and wishes of others. Above all . . . she was pure.[31]

One can see in this paragraph how easily an ethic of care might become misinterpreted as a form of slave ethics, and that is why it is important to analyze its components—to embed sympathy in empathy, to differentiate between self-sacrifice and the mutual gains of relatedness, to recognize the "difficult arts of family life" and consider how everyone should be educated to master them, to consider what it means to have a mind of one's own, to explore in some depth what it means to care in both private and public life. Religion, more often than not, has encouraged the myth of the Angel in the House. The task in care ethics is to retain and further develop the inclination to natural caring, but to do this in a way that does not perpetuate the subordination of women.

If reforming religion from within will not work, it has been suggested that restoring or reinventing Goddess religion might work for women.[32] Here again I agree with Phillips that this is not a promising course of action. First, we encounter a historical problem: there is little credible evidence that the status of women was better when goddesses were worshiped. Second, there is a risk of deposing one tyrant only to suffer under another. Without trying to restore the Great Goddess or to pretend that there is actually such a deity, it still might be useful to ask ourselves what sort of deity women might create in their own image. If we look at the Ten Commandments supposedly issued by the biblical male deity, we see a highly self-centered God; four of the commandments prescribe the duties of humans toward this God. Despite the claims of those who would post the Ten Commandments in public places because, ostensibly, they are the foundation of Western law, only three of the commandments appear in our law, and they were almost certainly in place before the creation of the commandments. What might our imaginary mother God issue by way of commandments? They would probably be cast as reminders, not commandments, and they might look like this:

Remember that I will always love you.

Love, protect, and gently instruct your children in the ways of caring.

Do not kill.

Do not rape.

Do not steal.

Do not bear false witness.

Be merciful to nonhuman beings.

Care for the earth and all of its life.

Have fun on Sunday.

Alas, there is no mother God, but happily there is no father God either. However, we learn something about ourselves by engaging in this exercise. What sort of men would invent the God of the Bible? Who profits from that invention? Why are so many people—both men and women—reluctant to consign this god to mythology? What purposes are served by preserving him?

It is often argued that religious communities help to sustain the basic morality of their members, and there is clearly some truth in this. When religious communities stand for the good, they are likely to play a positive role in encouraging decency and fellow-feeling. They may also protect their young from the various immoral practices that often surround and tempt them.[33] But religious communities are not always good, and when they are bad, they can exert enormous negative influence. Most theologians admit this possibility but insist that a religion can avoid this result by, as Charles Kimball puts it, remaining "true to its authentic sources."[34] I think Kimball is wrong on this. Not only is it difficult to prevent the perversion of authority discussed earlier; in most cases, it is the authentic sources that must be rejected. We, women and men, must work to create truly moral communities without the awful burden of traditional religion weighing us down. We might even discover a spirituality that inspires and comforts us without instilling fear, distrust, abject submission, hatred, and rigidity.

In the next chapter, we must do some more comparative work. Care ethics puts great emphasis on emotion and feeling. Where might we find philosophical agreement on this emphasis?

SIX

Emotions and Reason

In the last chapter, I attempted to distinguish care ethics from Confucianism, Christian agape, and more generally virtue ethics. Now we must look at the school of moral theory often thought to be the philosophical forebear of care ethics—moral sentimentalism. Michael Slote, in explaining the possibility of a care ethics more widely applicable than that described in feminist philosophy, remarks: "The ethics of care is historically rooted in the moral sentimentalism of Shaftesbury, Hutcheson, Hume, and Adam Smith, all of whom were men."[1] And Annette Baier has suggested that David Hume might be considered "the woman's philosopher."[2] There is certainly a historical debt to be acknowledged. However, just as we located both similarities and differences in the views treated in chapter 5, so we will find them again in moral sentimentalism.

Care theorists agree with moral sentimentalism on the primacy of emotion in moral life. We can also agree on matters concerning the connection between thinking and feeling, although the difference between natural caring and ethical caring raises some issues. Because I am attempting to develop care ethics from its origin in maternal instinct, I will have to say something about male/female differences in emotional life and how these differences might affect our views of the moral and social virtues. Care ethics is largely in agreement with moral sentimentalists that the social virtues play a central role in both moral and social life. However, in developing an ethical approach rooted in female experience, we may find significant differences on how the virtues develop and on which virtues are emphasized. Finally, again consulting female experience, we should have something to say about why and how to cultivate the moral sentiments.

Feminists are rightly wary of using the label "moral sentimentalism," because females have long been accused of being emotional and sentimental, as incapable of the principled reasoning required in ethics and morality. Although we want to make a central place for emotion and affect in moral life, we may prefer to avoid the language of sentimentalism.

MORAL SENTIMENTALISM

The third Earl of Shaftesbury is thought to be the first philosopher to use the term *moral sense*. He advanced the argument that human beings are naturally virtuous and that morality is not derived from religion. Indeed, religion depends on some already existing sense of morality; it does not invent morality. Care theorists can agree with Shaftesbury that morality is somehow natural, and some of us agree strongly that morality is not dependent on religion. Our concern is to explore this "naturalness." In what does it consist? Where does it arise? One answer to this is, of course, that it arises in maternal instinct and proceeds, with cognitive development, from instinctive caring to natural caring. I would not argue that "man is naturally virtuous," because we can reasonably pursue questions about the development of both virtues and vices. Something—some sense and the practices it demands—provides the setting in which virtues arise. Keep in mind that an alternative source of morality is self-interest, and I have not denied its salience. However, like care theorists, Shaftesbury was at pains to show that Hobbes's view of man's nature as basically selfish is not the only possibility. Self-interest can lead us to treat others well, so that in turn they will treat us well, and self-interest is probably the root of most concepts of justice. We might reasonably speak of *two* evolutionary paths to morality: maternal instinct and self-interest.

Francis Hutcheson analyzed the concept of *moral sense* in further detail. He identified the moral with that which pleases us and the immoral with that which displeases. His view was not equivalent to simple emotivism, however. For Hutcheson, it was not a matter of our saying, "Boo!" to some things and, "Hurrah!" to others. Hutcheson believed that God created human beings to be benevolent, and unless something goes badly wrong in their environment, most people are basically good. In other words, God made human beings who are pleased with what *should* please them. This is not only questionable but unsatisfying in that it leaves us with many unanswered questions. Why are some people pleased by things that displease others? Why are some things pleasing in moderation and displeasing when a certain line is passed? And where should that line be drawn?

David Hume agreed with Hutcheson and Shaftesbury that emotions or feelings are central to moral judgment. It is emotion, not reason, that motivates us to act. Reason, Hume argued, is attached to facts and judgments about facts; emotion controls moral judgment. We may discover truth and yet remain indifferent to it; it may have no effect on our conduct: "What is honourable, what is fair, what is becoming, what is noble, what is generous, takes possession of the heart, and animates us to embrace and maintain it. What is intelligible, what is evident, what is probable, what is true, procures only the cool assent of the understanding."[3] Hume argued that we make our moral judgments on the basis of whether a quality or an act is agreeable or disagreeable, useful or harmful. Like Hutcheson, he does not define the pleasant or agreeable in terms of idiosyncratic whims, but he does not follow Hutcheson in claiming that God has made people basically good. Rather, Hume asserts that human beings have a natural tendency toward benevolence—to be pained by another's pain, troubled by another's predicament, pleased by another's pleasure. Hume (probably an atheist) does not claim, however, that people are born good, and God plays no part in his argument. Instead, Hume says that human beings have access to all basic human emotions, including those that induce benevolence, and he then explores in some detail the effects of upbringing and social conventions in encouraging the emotions associated with benevolence.

For Hume, sympathy is the universal human emotion that links us together as individuals in particular circumstances and as human beings in a universal moral world. Alasdair MacIntyre claims that Hume was mistaken in this move, indeed that "'sympathy' as used by Hume and [Adam] Smith is the name of a philosophical fiction."[4] I think MacIntyre is wrong on this. Hume did his work under what MacIntyre considers a handicap; he rejected the notion of a divine purpose, or telos, for human life, and therefore he faced a difficulty in pronouncing on innate emotions or tendencies. But with the support of today's anthropological and psychological evolutionary theory, Hume would find exactly the support required for his claim to the innateness of sympathy in human beings, but that innateness is derived from experience shared with nonhuman animals, not from divine creation.[5] Sympathy is extended from its origin in maternal instinct to the natural caring we find in close groups, and it provides the motivation for the creation of ethical caring.

On careful examination, we can see that it is MacIntyre who faces the greater problem when he roots the virtues in tradition and practices. The virtues are, of course, described somewhat differently in different times and cultures, and some virtues do in fact arise within traditions and practices. The question that interests me in this book, however, is where the virtues that make traditions possible came from. MacIntyre depends heavily on the concepts of *tradition* and *practice*. But how did these concepts develop? In agreement with evolutionary anthropologists, I trace these developments to a rudimentary social structure that preceded the specifically human community. Further, I am concerned that the traditions and practices discussed by MacIntyre are located almost exclusively in male experience. No notice is taken of the virtues that arise directly in female experience or of the practices that—clearly articulated—might challenge the male ordering of virtues. Even in his generous references to Jane Austen, MacIntyre attributes the virtues she extols to her Christianity, not to her experience as a woman.

MacIntyre recognizes that both emotion and reason are required in moral life. Although emotions drive action and moral judgment, reason is clearly involved, and this is the case in care theory as well. Hume notes, for example, that justice does not arise as a simple instinct. Rather, we reflect on the ways in which justice contributes to both happiness and utility, and we make a commitment to it. We study the concept, revise it when necessary, and build our laws and conventions on it as a foundation. I said earlier, however, that justice may more naturally arise from the basic instinct of self-interest than from the sympathy associated with caring, but care ethics may contribute substantially to a modification of justice. I'll return to that discussion in chapter 7.

For care theorists, it is especially interesting to note how Hume emphasizes the social virtues. Like care theorists, he is interested in moral *life*, not just principles and judgments. He gives substantial attention to virtues that are pleasant or useful to ourselves and to others. The social conventions that make our lives pleasant are much to be admired. All of us know, on reflection, that it is easier to behave morally toward a polite, cheerful person than toward a rude, difficult curmudgeon. Care theory recognizes and builds on this observation when it describes a role for the cared-for in maintaining caring relations. Thus, if we follow Hume (and in this respect care theory does), we will put appropriate emphasis on the social graces and help our children understand which conventions should be accepted for their contribution to moral life and which can be ignored without damaging the moral network.

Hume did not closely analyze attention, but concern with it is implied in his discussion of the qualities we admire most in others: "No qualities are more intitled to the general good-will and approbation of mankind than benevolence and humanity, friendship and gratitude, natural affection and public spirit, or whatever proceeds from a tender sympathy with others."[6] Here we might recall that sympathy sometimes precedes focused attention; that is, our attention is occasionally seized by the immediate sympathy or pity we feel for a suffering person or nonhuman creature. As Hume pointed out, it is a rare human being who is never moved by sympathy. However, in building a moral life, we cannot depend on the dramatic appeal of extreme situations. We must cultivate a form of attention that keeps open the connection between perception and sympathy. In doing so, we become more vulnerable, for what hurts another will also hurt us. Some people—perhaps only unconsciously influenced by this knowledge—construct a wall between their perception and the feelings that should follow. They may achieve empathic accuracy, but they are not moved by sympathy. Caring depends on a form of attention that requires continuous interaction between the cognition that assesses needs and the emotion that moves a carer to respond to them.

In the last section of this chapter, we will return to an appreciative examination of Hume's analysis of social virtues—virtues of central importance to care ethics, but first I want to say more about the interaction of cognition and emotion.

THINKING AND FEELING

Emotion in human life is clearly prior to language and the forms of cognition associated with language. Humans share an impressive range of emotions with other mammals, and signs of emotion endear many animals to us. It is no longer assumed that an observer who attributes emotion or personal idiosyncrasies to nonhuman animals is guilty of anthropomorphizing.[7] This literature is accompanied by works defending the notion that we have some moral obligations to animals—not because they think as we do, but because they can feel as we do. Jeremy Bentham put it this way: "The question is not, Can they *reason?* Nor Can they *talk?* But, Can they *suffer?*"[8] I cannot undertake a survey of either branch of this fascinating literature here. Its importance for present purposes is to support the claim that morality is derived from emotions that precede consciously human relations.

Jonathan Turner argues that natural selection for emotional sensitivity contributed to the social characteristics of human beings and thus to their survival in the savannah environment. He entertains the possibility that maternal instinct might have contributed something to the development of human sociality. He asks, for example, "Could selection extend the mother-infant bond to include males, thereby creating cohesive groups?"[9] He ignores the fact that half of our infants are male and, therefore, automatically influenced to some degree by the mother's instinct to care. He also overlooks the role that might have been played by female-male relations in favoring selection of a "more emotional animal"—the human being. We do not have to claim that the maternal instinct and its associated emotions are the only source of morality to appreciate their likely contribution.

In contradiction to Turner's contention that humans are descended from "lightly social" hominids, Frans De Waal argues for a more social ancestry: "There never was a point at which we became social: descended from highly social ancestors—a long line of monkeys and apes—we have been group-living forever. . . . We come from a long lineage of hierarchical animals for which life in groups is not an option but a survival strategy."[10] De Waal uses these evolutionary facts effectively to discredit political theories that posit self-interested individuals who enter voluntary, contractual arrangements to establish societies. But he too brushes aside the maternal factor too quickly: "While the primacy of connectedness is naturally understood by women— perhaps because mammalian females with caring tendencies have outreproduced those without for 180 million years—it applies equally to men."[11] But, as we discussed earlier, male-female differences persist, and human survival may depend on understanding and overcoming some of these differences. How exactly does the primacy of connectedness apply to males?

In addition to the evolutionary evidence that links social affects and emotion to morality, there is also information on contemporary human behavior that supports the link. In their study of non-Jewish rescuers of Jews during the Holocaust, Pearl and Samuel Oliner found that relatively few rescuers cited principles as the reason for their engaging in rescue activities. Many acted from allegiance to community norms—"this is what our community expects of us," and many others responded directly through sympathy. The Oliners comment:

> In a culture that values individualism and rational thought most highly, a morality rooted in autonomy is considered most praiseworthy. Those who behave correctly—ethically, in fact—but do so in compliance with social norms or standards set by individuals or groups close to them or because of empathic arousal are presumed to be in some way morally deficient. That few individuals behave virtu-

ously because of autonomous contemplation of abstract principles—a finding that has been reiterated in numerous studies including Adorno's and our own—has not deterred advocates of independent moral reasoning from advancing it as the most morally admirable style.[12]

They go on to observe that the very rarity of principled thinking may raise its value in the eyes of some observers. We admire selfless, moral heroes. But, they add, "This is also a dispiriting view, for if humankind is dependent on only a few autonomously principled people, then the future is bleak indeed."[13] We should note also that many Nazis and others who engaged in cruel and criminal activities pointed to principles to justify their conduct. The corruption of Kant's concept of duty has been widely recognized. Similarly, adherence to community norms depends for its goodness on the moral goodness of the community. A community gone bad will support its members in questionable, even evil behavior.[14]

None of this is to say that reasoning is not used in responding empathically. It is disheartening to note the comments of some critics of *Caring*. Martha Nussbaum, for example, accuses me of recommending "thoughtless giving." She writes: "Nel Noddings holds that women's experience of mothering reveals a rich terrain of emotional experience into which judgment and appraisal do not and should not enter. . . . Unless we give ourselves away to others without asking questions, we have not behaved in a fully moral way."[15] This would be a devastating criticism if I had said such a thing, but I did not. Nussbaum's misunderstanding seems to have arisen from her concentration on one paragraph in which I described *an occasion* of emotion that arose without any conscious reflection. Here is the fragment quoted by Nussbaum: "There is the joy that unaccountably floods over me as I walk into the house and see my daughter asleep on the sofa. She is exhausted from basketball playing, and her hair lies curled on a damp forehead. The joy I feel is immediate. . . . There is a feeling of connectedness in my joy, but no awareness of a particular belief and, certainly, no conscious assessment."[16] Obviously, this paragraph says nothing about "thoughtless giving," nor does it recommend that "women should simply give themselves away without demanding a just distribution of resources."[17] Indeed, I do not make such a recommendation anywhere in my writing on care ethics. Drawing on Buber's work, I discuss the desirable mutuality properly expected between competent adults in equal relations such as marriage and friendship. Within a relationship, I may be carer in one episode, cared-for in another. This mutuality is what makes life pleasant and good; it maintains the environment of natural caring. Although I point out regularly that the pair (carer, cared-for) may be (male, female), (female, male), (male, male), or (female, female), I do speak of the carer as "she" and the cared-for as "he," and that choice of linguistic convenience may have contributed to misunderstanding.

Nussbaum was distressed by my description of a noncognitive joy that occasionally floods over us, but she overlooked the project of the chapter in which the story appears. I was asking questions of this sort: Does emotion always have an object? Should we distinguish between joy as emotion and joy as reflective affect? Is emotion a purely physiological response? Is cognition always involved in emotion? Is it sometimes involved? In my discussion of the story cited by Nussbaum, I noted that, although my daughter seemed to be the object of my momentary emotion, the recognition of relation was even more directly responsible for my response. This is my child, and we've been connected since her birth (even before that) by the most tender bonds. In many, perhaps most, occasions of emotion, the relation or situation necessarily colors the nature and intensity of feeling. In battle, for example, a soldier may feel blind rage at an individual representative of the enemy, but it is the setting or situation that creates fear and fury. In another situation, this particular "enemy" might be a comrade on the playing field. Nor do I think I was wrong to describe such moments as lacking awareness of a particular belief or any conscious assessment. All cognitive awareness and conscious assessment have been somehow built into the situation.

I should say a bit more about the lack of conscious cognitive assessment in many situations. In settings characterized by natural caring, there is no need for continual analysis and assessment; most evaluation is almost automatic—coming "prepackaged" as it were. Only an unusual occurrence or perception (accurate or inaccurate) triggers analysis. In asserting that I was wrong to claim "no awareness of a particular belief . . . no conscious assessment," Nussbaum writes that I must have held the beliefs that my daughter was asleep, not dead, and that it was indeed my daughter, not a burglar, on the couch.[18] But this is somewhat silly. In familiar settings and contexts, we do not ask ourselves such questions. Rather, our emotional channels are comfortably open, and feeling arises directly. If my daughter had been (unexpectedly) dressed in some odd Halloween costume with her back toward me, I might indeed have asked, Who is this? It is not that cognition is not at all involved in most everyday instances of emotion, but assessment has become unnecessary—routinized.

Besides the analysis of emotion and feeling, I had another purpose in that chapter, and that was to pose a counterpoint to Sartre's identification of anguish as the basic human affect—one accompanying the realization of human aloneness and absolute freedom to choose. When we recognize relation as ontologically basic to human life, we do not rid ourselves entirely of anguish or of a responsibility to choose (within limits), but we do open our minds and hearts to those affects that accompany a realization of relatedness. Joy as emotion floods us now and then. More often, joy as affect or its softer variant, contentment, may pervade our lives. It is this, in part, that we seek in natural caring.

Recognition of life-affirming feelings underscores a point at the center of care ethics—the primacy of natural caring. On this, Nussbaum's misunderstanding is very serious. I have *not* argued that natural caring is *morally* superior to ethical caring. I have argued that natural caring precedes and establishes a model for ethical caring. Natural caring also provides motivation for ethical caring, because natural caring is our *preferred* social condition. Ethical caring, as a form of formal morality, serves natural caring. In *Caring*, I wrote: "The relation of natural caring will be identified as the human condition toward which we long and strive, and it is our longing for caring—to be in that special relation—that provides the motivation for us to be moral. We want to be *moral* in order to remain in the caring relation and to enhance the ideal of ourselves as one-caring."[19] As we discussed earlier, this relation may occur as a mere encounter—address and response between strangers. We all prefer to live in a society where such encounters are friendly and helpful. Or the relation may take the form of an episode—a longer interaction that is complete in itself or part of an ongoing set of interactions within a stable relationship. Continuous relationships such as parent-child, sibling-sibling, husband-wife, partner-partner, or friend-friend are rightly labeled *caring* if most of their encounters and episodes are caring. In all equal relationships, the members take turns as carer and cared-for.

Natural caring is clearly emotion-based. We have some neurologically based capacity for reading the emotional state, needs, and intentions of others, and with appropriate guidance, we can bring our empathic capacity to a high level.[20] The roots of this capacity can be found in our nonhuman ancestors. De Waal remarks: "A . . . debate pitting reason against emotion has been raging regarding the origin of morality, a hallmark of human society. One school views morality as a cultural innovation achieved by our species alone. This school does not see moral tendencies as part and parcel of human nature. Our ancestors, it claims, became moral by choice. The second school, in contrast, views morality as a direct outgrowth of the social instincts that we share with other animals."[21] Largely in agreement with the second school, I argue that (formal) morality properly construed serves the purpose of preserving and enhancing the culture of natural caring. Because social virtues develop from social instincts, we must work to cultivate social virtues, and because moral thinking is rooted in emotion, we must cultivate the moral sentiments.

Again, I am not arguing that reason is not involved in either natural caring or ethical caring. Instrumental reasoning is obviously involved in natural caring. Although I do not need to refer to a principle or ask what my community expects of me in this situation—I know that I will respond in some positive way—I do have to consider exactly what I will do, and I may have to spend considerable time in assessing the likely effects of my well-intended choices. But that whole process will be colored by what I feel. My reasoning may well modify my feelings, but the motive energy to act will come from my (modified) feelings.

In ethical caring, there is a larger role for reasoning. I may have to overcome an initial reluctance to act on behalf of the other, or I may feel unsure about the desirability of maintaining or establishing a caring relation with this person. In such cases, I turn to my own ethical ideal of caring to answer the question, How should I respond? But clearly, that ethical ideal—under continuous construction and evaluation—is affect-loaded and dependent on the past cultivation of social virtues and moral sentiments.

EMOTIONS, VIRTUE, AND GENDER

Before exploring the role of social virtues and moral sentiments in a happy, moral life, we need to say more about gender differences in the experience and expression of emotion. As I have pointed out, it has long been said that men are rational and women emotional. Reasonably impartial observation should disabuse anyone of this notion. Recently, a photo of a successful male athlete appeared in all the major newspapers. Having won an outstanding victory in a tennis match, he was lying on his back, kicking both feet in the air like a happy infant. Why not? He was ecstatically happy. We see such exhibitions almost daily in the sports world. And we regularly see angry mobs (mostly male) shouting, waving fists and guns, and wreaking havoc in various parts of the world. Surely such scenes should make us rethink the faulty notion that women are more emotional than men.

It is likely, however, that females and males experience some emotions more often and more intensely than the other sex. For the survival of their infants, females have developed a heightened capacity for sympathy, and for their own survival they have allowed emotions that might fall into the category of "wariness" to be converted to something like the sympathy that triggers motivational displacement. Watchfulness and wariness are regularly connected to sympathy-like feelings that promote acting-for-the-other. Clearly, there is a cognitive element in this conversion. Females must often behave "as if" they are sympathetic when their truer feeling is fear or anxiousness. It seems reasonable, too, to suggest that this ability to misrepresent their true emotion would contribute to a certain emotional obtuseness in males. As a result, males might have difficulty achieving empathic accuracy.

In contrast, males more often experience and express anger and aggression. Indeed, these emotions—as long as they are not directed at the "wrong" objects and do not rage out of control—have been respected as signs of "real men." The development of empathic skills in males has probably been encouraged by the need to detect fear, anger, and aggression in other males. Male reading of the other in strange or unfriendly surroundings is less likely to induce sympathy, the primary ingredient in female empathy.

Both males and females experience fear, but the female's may be disguised as wariness and directed toward activities that placate the male and increase his sense of mastery. The male's fear may manifest itself in bravado, belligerence, or incautious aggression.

Societal judgments on emotion have reinforced evolutionary gender differences. It is more acceptable for women than for men to break down in tears. In males, wild exuberance in victory is thought to be normal; in women, it is unseemly. Even assertiveness is judged differently—desirable in males, resented in females. Fear is recognized as normal, even prudent, in both sexes, but *flight* is more acceptable for women, and *fight* for men. The rules of warfare have reinforced this judgment. Men who flee the battlefield are held in contempt, sometimes legally killed. Until recently, sexual prowess has been admired in men and distrusted or shamed in women. These differences, often the material of cartoons and comedies, have played and continue to play a major role in the continuing subordination of women. Although it is politically incorrect to declare outright that women are inferior to men, the genderized valuation of attributes succeeds in maintaining a structure that keeps the belief intact, if muffled.

A historical look at the virtues and how they have been taught yields another bit of evidence on the devaluation of female experience. The Greeks put emphasis on courage, defined primarily as the courage of the warrior. Standing with courage at the apex of the virtues was wisdom. For the Stoics, courage guided and fearlessly sought wisdom. Later, for Thomas Aquinas, wisdom (reason) guided courage.[22] Both courage and wisdom have long been associated with *manliness*, a strategy that either denies these virtues to women or defines those females who possess them as somehow unnatural.

More subtly, in the early twentieth century, the Character Development League listed thirty-one virtues to be taught in schools.[23] It is interesting to note that many of the virtues thought appropriate for women—obedience, truthfulness, unselfishness, sympathy, usefulness, and patience—appear in the first ten, those to be taught earliest. Sympathy is said to require action, and that in turn requires "consecration to duty," a move that may or may not meet the needs of one receiving "sympathy." The more masculine virtues—fortitude, courage, self-reliance, justice, ambition, and heroism—come later.

Relatively little has been written on female virtues from the perspective of women. Greek philosophers (all male) extolled the feminine virtue of temperance, or control of passions, and later philosophers have pointed to the virtues mentioned above. But if we apply philosophical thinking to female experience, what virtue might we name as the equivalent of the male *courage?* That is, what female virtue might stand at the top of a female hierarchy of virtues?

I think it would probably be *sympathetic attention*, but it has no simple name as yet. I noted earlier that Simone Weil discussed receptive attention, but her view looks beyond the particular human being who addresses us to some divine image of humanity. It is heavily influenced by belief in the male deity. Iris Murdoch's position is much closer to that of caring. It connects directly to the living other who has needs and to a moral subject capable of "looking lovingly."

Sympathetic attention is a complex virtue. It springs from maternal instinct, and so has an evolutionary stamp of approval. But it involves skill, and that skill can be honed to a high level of competence. As I pointed out earlier, that skill might be called empathic accuracy, provided we understand that in care ethics empathy necessarily includes sympathy. Sympathetic attention requires more than feeling and skill. As a virtue, it requires *commitment*. A person who regularly responds with sympathetic attention has committed herself to keeping open the channels from perception to feeling to motivational displacement. The carer has cultivated the virtue of sympathetic attention, and she accepts the vulnerability that accompanies exercise of this virtue.

A care theorist, however, is somewhat wary of calling sympathetic attention a virtue. Sympathetic attention does not function as most virtues do. It directs us away from ourselves as admirable characters and toward the one who addresses us. We are more concerned with the *relation* than we are with our own status as virtuous persons. This attitude is clearly different from that in Aristotelian virtue ethics or in Confucianism. It might help to invoke another contrast. In a liberal, neo-Kantian view, John Rawls writes: "The virtues are sentiments, that is, related families of dispositions and propensities regulated by a higher-order desire, in this case a desire to act from the corresponding moral principles."[24] Neither a virtue theorist nor a care ethicist would accept this definition. Both accept principles as guides to everyday conduct, but both distrust principles when moral life gets really sticky. Then we must turn to something continually cultivated, treasured, and maintained within ourselves. The care theorist is even a little wary of this, because we know that ethical caring—precious as it is—is always risky and best employed to restore natural caring. In natural caring, our commitment is to the living other, not to a principle.

Sympathetic attention is accompanied by other attributes and habits that describe a caring person. In long lists of virtues, they rarely appear; indeed, some have no simple names. Frugality does appear on many lists, and it has figured prominently in female experience. However, in many discussions, it is used as a foil to ward against stinginess, a vice characteristic of exaggerated frugality. Careful examination of frugality in female experience exposes the important fact that frugality was, in a sense, the main way women had of earning a living. Many women in my mother's generation and earlier felt rightly that they provided the equivalent of a paid salary through the exercise of frugality.

But I am more concerned with the sort of attributes and habits identified by Sara Ruddick: preservative love, holding, staying with, fostering growth. Of "holding," Ruddick writes: "Holding is a way of seeing with an eye toward maintaining the minimal harmony, material resources, and skills necessary for sustaining a child in safety."[25] This is another way of talking about natural caring in the context of childrearing and how to encourage it throughout the web of care. Notice that "holding" is not the personal possession of a moral agent but can only be exercised in relation, and it is guided by the needs of the cared-for. We hold young children more closely than older children, but women often take responsibility for holding at every stage of life. For example, we exercise "holding" in the last days of our elderly parents. Sympathetic attention guides the exercise of holding and preserving life. Similarly, it guides the other two great maternal interests identified by Ruddick: fostering growth and shaping an acceptable child.

Is "holding" a virtue? It is not an explicitly defined activity like cooking or writing. It is not a virtue in the Rawlsian sense—a sentiment regulated by a desire to act on the prescribed moral principles. Nor is it, as MacIntyre would have it, "an acquired human quality the possession and exercise of which tends to enable us to achieve those goods which are internal to practices and the lack of which effectively prevents us from achieving any such goods."[26] On first hearing, there seems to be a resemblance between MacIntyre's definition and the female experience of holding. But reading on, we see an emphasis on standards established by authorities in the practice, and such standards do not play a significant role in holding or, more generally, in caring. We do not compare ourselves to other carers; we are not in a competition. Our specific acts and our evaluation of success come from the response of the cared-for. This is not to say that there are no people who, in professing to care, compare themselves with other "carers," give too much attention to authorities, and continually seek credit for their efforts. But these people have missed the point of caring and often fail to establish caring relations.

We do not know how to classify holding, preserving, staying with, conserving, and fostering—clusters of ill-defined activities central to mothering, nursing, and the best teaching. We need a whole new way of talking about activities—ways of being in the world—that are inherently relational and deeply embedded in female experience.

SOCIAL VIRTUES IN MORAL LIFE

There are, of course, social virtues that contribute positively to moral life. David Hume stands out as a philosopher who emphasized the social virtues, among them cheerfulness, politeness, wit, modesty, decency, and cleanliness.[27] It is easy to agree with

Hume on this. All we have to do is to ask ourselves how much we enjoy being around people who are glum, rude, dull, boastful, indecent, or dirty. Lack of social virtues strains the relations of natural caring. Would-be carers must call upon ethical caring again and again in their interactions with such people, whereas those who exhibit social virtues make moral response relatively easy.

Looking at the contributions of the cared-for, we recall that these responses, too, support natural caring. In some cases—that of the responsive infant, for example—we see pre-social virtues. The responsive infant may easily learn to be polite, modest, and decent. It is not only easier to maintain natural caring in relations with people who are socially responsive; it is also easier to teach responsive children something about the social graces and, thereby, further strengthen the whole web of relations.

Hume made his judgments about the social virtues on the basis of pleasure (to ourselves and to others) and utility. I make mine on the basis of their contributions to natural caring. I would add to those discussed by Hume such social descriptors as hospitable, nonargumentative, attentive (which is both a social and moral virtue), and adaptable. A person who is both attentive and adaptable may have a sardonic sense of humor, but she will know when to employ it, when to moderate it, and when to squelch it entirely.

Hume has critical words for virtues often associated with religious life: "celibacy, fasting, penance, mortification, self-denial, humility, silence, solitude, and the whole train of monkish virtues."[28] These give us no pleasure, nor do they have utility. Most care theorists, I think, would soften the criticism of silence and solitude for, indulged moderately, they help to restore our commitment to both caring and inner tranquility. But in general Hume is right to reject these "virtues." He may go a bit too far when he says: "A gloomy, hair-brained enthusiast, after his death, may have a place in the calendar; but will scarcely ever be admitted, when alive, into intimacy and society, except by those who are as delirious and dismal as himself."[29] Largely in agreement with Hume, Daniel Dennett criticizes the monkish tendencies from a moral perspective: "There are many people who quite innocently and sincerely believe that if they are earnest in attending to their own personal 'spiritual' needs, this amounts to living a morally good life."[30] Referring to contemplative monks who "devote most of their waking hours to the purification of their souls, and the rest to the maintenance of the contemplative lifestyle to which they have become accustomed," he asks, "In what way, exactly, are they morally superior to people who devote their lives to improving their stamp collection or their golf swing?"[31]

The great contribution of the social virtues in care theory is to support and enhance natural caring. But within domains characterized by natural caring, we are faced continually with demands or requests to listen to and respond to needs. The social virtues make the task easier, but it remains a major task within caring relations, and it has long been an extremely difficult topic for philosophers, economists, and other theoreticians. We turn to that topic next.

SEVEN

Needs, Wants, and Interests

Care ethics is oriented to needs rather than rights. This orientation seems exactly right in the context of families and small communities, but it becomes more difficult to sustain in larger settings. Indeed some philosophers have argued that the concept of needs is too complex to employ usefully in policy decisions. I mentioned earlier in my very brief discussion of a care-driven approach to justice that care ethics is centrally concerned with needs. In the same chapter, I explored some of the difficulties we face in caring at a distance and deciding which obligations are individual and which collective. Now we should discuss these matters further. I'll start the chapter with an examination of some of the problems involved in working with needs and then move on to a discussion of basic needs, wants, and interests.

WORKING WITH NEEDS

Theorists working with the concept of needs encounter the problem of identifying and interpreting needs.[1] The problem is especially acute for those who start at an abstract level and seek well-defined principles on which to base social/political policy. Care ethics is in a better starting position, because we do not start with an abstract notion of need. We begin our thinking within the concrete conditions of natural caring. It is not a matter of identifying needs and then employing care to meet them. *Caring precedes the identification of needs.* Even some advocates of care ethics overlook this basic fact. Joan Tronto, for example, writes: "Since caring rests upon the satisfaction of needs for care, the problem of determining *which needs* should be met shows that the care ethic is not individualistic, but must be situated in the broader moral context. Obviously, a theory of justice is necessary to discern among more and less urgent needs."[2] But this is *not* obvious. There may be another way to approach the problem and, as Tronto does point out, the required theory of justice (if it is required) may be different from the ones we are familiar with in liberal philosophy. Again, although Tronto places attentiveness first among four elements of care, she moves quickly to care defined as meeting needs—the form of caring that I have called *caregiving*.[3] But there is more to care than caregiving. If there is a basic need addressed by care ethics, it is the need to be heard, recognized. In the conditions of natural caring, each human being is comfortably aware that *if* a need arises, someone in the circle of care will respond. This assurance of response characterizes natural caring. A particular need may or may not be met, but it will receive a sympathetic hearing.

It is impossible to overemphasize this point. The conditions of natural caring establish the best climate for the identification of needs. In good homes and small communities, most needs are identified person-to-person. The needs are not always met, but they are heard, and explanations are offered when the needs cannot be met. In many good homes, families work together to sort out which needs must be met, which may be met if resources can be found, and which must be deferred or left unmet. Part of this process involves differentiating among needs and wants, and we'll discuss this process in more detail in each of the following sections.

A second difficulty arises when we start with the *concept* of needs and try to proceed at an abstract, conceptual level. David Braybrooke identifies three points at which the concept breaks down or encounters a bottomless pit of needs.[4] The difficulties are real, but I think liberal philosophers are looking in the wrong direction to solve them. In the first case—that of embarrassing or ignoble needs (e.g., addiction), philosophers may tie themselves in knots trying to decide whether such needs should be admitted to the list of needs and, if they are listed, what priority should be assigned to them. Surely, when this happens, we should recognize a turning point and start over again from a real situation. How do we respond when a family member has an addiction? How do we respond in schools when a student has a seemingly overwhelming need to hurt others? I'll look at this problem in the last section of this chapter. The other two breakdowns identified by Braybrooke—the bottomless pit of medical care and application of the concept of needs on a world scale— will be addressed in the next two sections.

A third difficulty is located in the treatment of needs as hierarchical. Abraham Maslow, for example, suggested that needs are organized in a hierarchy: physiological needs, safety, love, self-esteem, and self-actualization.[5] It is almost certainly true that a person whose physiological needs are unmet will not have much time or energy to pursue self-actualization. Indeed whole populations may be ensnared in a grinding effort to achieve a minimal level of physiological satisfaction.

But that does not mean that the needs do not coexist. No mother would approach her child's needs this way: first, I'll feed you, then make sure you're safe, then give you lots of love. The needs, if they are there at all, are all there at once. We might raise questions about the last two of Maslow's needs, especially about their interpretation. Not all cultures put a high value on individual self-esteem, and the need for self-actualization may be relatively rare. However, they are clearly important in contemporary Western culture, and the demand to meet them falls into the category Ruddick calls "fostering growth." I'll say more about this too.

BASIC NEEDS

It is obvious that all human beings need food, water, and shelter. It is these needs, as we saw earlier, that Peter Singer urges us to meet worldwide. Although we know that these needs are common to all of humanity, we are not always sure about their urgency, and we can make errors in how to respond to them. When people in the developed world hear that other people are hungry—as they have been for some time in Darfur—we rightly respond with food. But in some cases, if hungry people were settled in safety, they could meet many of their own needs for food. It may be a mistake to respond with food and neglect the equally great needs for safety and self-respect. In addition, by encouraging caregivers to distribute food in dangerous places, we put them at risk as well. An emphatic and convincing reply to this line of argument is to point out that people dead or dying of starvation no longer need either shelter or self-respect. So of course, we must try to provide them with food, but I am arguing that providing food is not enough. And if it were not for saving their children, many women would prefer to die than to live as they must to secure food.

In an environment of natural caring, physiological and emotional needs are addressed simultaneously. Indeed, history has recorded situations when courageous human beings have sustained one another emotionally although they could do little to meet even minimal physiological needs.[6] In chapter 8, we'll see that the moral failure to respond to fellow human beings *as persons* with needs that should at least be recognized induces emotional suffering that is harder to endure than physical suffering.

Ideally—perhaps I should say *realistically*—the first step in identifying and meeting needs is to establish the conditions of natural caring. When these conditions are in place, it is easier to respond to natural disasters such as earthquakes and floods, and good-hearted people contribute generously to meet the obvious needs. When such communities of caring have not been established, when strangers try to meet assumed needs, well-intentioned helpers may do foolish things—flooding a devastated region with clothing when it needs building materials, collecting acres of mobile homes unsuitable for the terrain in which people are homeless. We have learned, too, that many people feel a need to preserve the lives of their pets and would prefer to suffer days of deprivation rather than forsake the animals to whom they have made a commitment. In conditions of natural caring, we listen to one another, recognize the emotions others are experiencing, negotiate, and respond as positively as the situation and resources permit.

David Braybrooke identifies two basic ways of considering needs. On the one hand, some thinkers urge us to take the attitude of World Citizen and respond to needs on the basis of priority. Others advise us to adopt an attitude of Charity Begins at Home.[7] Braybrooke does not invoke an evolutionary argument, but he does acknowledge that our traditional moral and conceptual structures favor Charity Begins at Home, and I have already argued that it is a mistake to overlook "how we are" in our moral theorizing. In emergency situations, we will choose to stand with our own. This recognition suggests a thoughtful plan of action to prevent situations that force such choices.

Another complicated problem arises in the World Citizen approach. If we decide that basic needs should be met worldwide as a first priority and only then should resources be used for various luxuries closer to home, we run into problems associated with relative deprivation. There are people in the United States, for example, who are certainly better off than the poorest in underdeveloped countries but are nevertheless severely deprived in comparison with the middle range of families in their own country. It is not obvious where the line should be drawn in withholding aid to our own poor in order to support those a world away. People of good will may wish they could adopt the stance of World Citizen, but realistically they believe that we must work our way carefully to that position.

A realistic beginning demands a fundamental change in our approach to the whole concept of needs. Instead of compiling lists of needs in defensible priority and engaging in efforts to rescue Utilitarian thinking from the disasters it may inevitably produce, we might concentrate on extending existing communities of care to somewhat larger ones and on identifying on-site centers of natural caring in which we—as strangers—might responsibly participate. Needs are best identified and interpreted within circles of caring.

It is, however, unlikely that we will soon build communities of natural caring that connect nations worldwide. Actualization of such a vision requires not only providing for the basic needs of people as quickly as possible but also for the slow, patient work of establishing extended circles of natural caring. Sometimes, it is just a matter of recognizing and respecting those circles—communities already in place—that can work with outside providers to identify and meet needs. We must work to be invited into these circles. To get started, providers (nations, agencies) have to recognize the need for collective action. The problems will not be solved by individual, voluntary contributions.

Why is it that a wealthy country such as the United States gives such a small proportion of its wealth to those in need? Other developed nations give, proportionally, much more. One reason may be that the United States retains an exaggerated commitment to individualism. Sharing is not thought of as a collective way of life. Rather, providing for others is regarded as a personal virtue, one that contributes to both the welfare of the "needy" and to the salvation of the giver. It may be that allegiance to dogmatic religion actually impedes the movement toward collective action. We think of those to whom we give as somehow inferior, not just unfortunate—especially if their religion differs from ours. They have not achieved self-sufficiency.

And too often we identify our giving with our own status as members of a religious institution. The very idea of World Citizen is suspect in many religious circles, and those who adopt it may be criticized as unpatriotic. Recall the case of Thomas Paine who declared himself a World Citizen and his religion "to do good." Years later, Theodore Roosevelt heaped scorn on Paine as a "filthy little atheist."[8] People are praised for giving to the unfortunate but rarely for joining them as equals in a cooperative project.

Human beings have not made great progress in this area. In the next chapter, we'll discuss the powerful ways in which allegiance to a national identity are promoted. Religious institutions are prominent among the organizations that maintain this allegiance. This is not to say that religious institutions are insensitive to the plight of those who are hungry and oppressed. On the contrary, they are often active in trying to relieve suffering. But there is always the lingering concern with personal salvation—that of the victims and that of the givers. The good that is done—and it is not insignificant—is partly offset by the motive of self-interest. Those nations least influenced by religion seem to be most advanced in producing the attitude of world citizenship.

An objection to this claim arises immediately. What about godless communism and the fascisms so prominent in the middle of the twentieth century? Surely it was not religion that turned them away from world citizenship. Quite right. The claim is that *any* ideology that requires specific beliefs of its members and rejects all others as unsaved or somehow outside the fold—any such ideology—works against the development of world citizenship.

It has been charged that care ethics—especially as I have described it—encourages parochialism and discourages the attitude of world citizenship.[9] This is a serious charge, and I acknowledge that, in insisting that we must respond to those who address us directly, there is a risk that the daily occurrences of address and response will fill our lives, leaving little energy and few resources to meet the real needs of those at a distance. But it is not the ethic of care that produces this result. Caring stretches individuals beyond their capacity to respond, and the appropriate remedy is committed collective action. A care-driven approach to justice would permit those who "care about" to contribute collectively to funds that could be used to create environments in which "caring for" should flourish, and many women would vote positively for an increase in taxes specifically designated for the worldwide relief of poverty. Such a move would in effect take the World Citizen stance by directing money "off the top" to meet the most crucial needs. Studies have consistently shown that women are more likely than men to share what they earn, and individuals committed to care ethics would be relieved of the anxiety-producing need to reduce or sharply curtail the resources directed at those closest to them. Their net income would be somewhat reduced, and they could give even more from what is left if they wished to do so, but an ethical burden would be lifted by this collective commitment. At present, charitable giving outweighs government support by more than 10 billion. It is doubtful that individuals and families will decide on giving even more. However, those committed to an ethic of care may well accept a lower net income to relieve world poverty.

Many writers have noted that the caregiving most longed for and admired has been provided by women in the context of homes and small communities. To encourage this type of caring, it is argued, is to encourage parochialism—to confine care to our inner circles. But the women who gave this care for centuries had little power in the public world, and I argue that it is this powerlessness, not caring, that maintains parochialism. We care where we *can* care. Remember, too, that the money collected by a care-driven approach to justice will not in itself solve the problem. That money will have to be used in a way that establishes circles in which caring-for can flourish.

It is necessary that the resources to meet basic needs be collected and distributed by collective, not individual, agencies. When we look at yearly surveys of social attitudes, we see that women—the traditional caregivers—express significantly greater concern than men about poverty, hunger, child welfare, and the unmet needs of many groups. We are biologically constituted to care first for those closest to us, but that does not mean that our concern ends there. My guess is that women would overwhelmingly approve the appropriation of a reasonable amount of our nation's wealth to a fund committed to meet basic needs all over the world.

But while funds must be provided by the collective, caring-in-action must reinstate person-to-person relations. Small institutions designed to identify and meet needs should be established throughout the world. The orientation of these institutions should be toward human flourishing; that is, they should be charged and enabled to consider needs holistically.

Full-service schools represent an example of the institutions I have in mind. In the United States, full-service schools in our inner cities would provide not only academic instruction but also medical and dental services, nutritious meals, day care for preschool children and infants, social services, and legal counsel. Further, the activities associated with the academic role of schools would be generously expanded to the arts, democratically organized clubs, cross-age reading and discussion groups, and opportunities for community service. The establishment of full-service schools in some of our own cities would provide a model for similar institutions worldwide that would be funded by designated tax moneys from participating nations. A few such schools already exist.

Even in near-ideal situations, there would be competition among needs—just as there is in families. However, the response to competing needs would be worked out on site—just as it is in families. We will not meet worldwide needs by making lists of priorities and adhering stubbornly to them, nor will it help much to establish arbitrary minimal standards. We need a network that can respond to conditions in particular places, and we should give more attention to an education of citizens designed to develop empathy, negotiating skills, generosity of spirit, and commitment to caring.

In addition to physiological and safety needs, human beings have another need that is rarely mentioned. We seem to have a basic need to have at least a few of our *wants* satisfied. Most parents recognize this need in their children. Wise parents watch for signs that a want persists. Interests and wants interact over time, and when a sustained interest induces a relevant want, parents try to meet it. They may even say, "Patty *needs* a microscope," in recognition of Patty's obvious interest in ocean water, puddles, pond water, and tap water, whereas they would not claim a *need* for all of their children to have microscopes.

WANTS, INTERESTS, AND NEEDS

In the best homes, the need to have some wants satisfied is given respectful attention. Parents recognize physiological needs, of course, and they also attend to the needs associated with shaping an acceptable child. The latter needs differ across cultures and within cultures, across religious groups and various traditions. We might say that these needs are *imposed* on children, for although they have a need to be accepted, they have to learn how to behave in order to achieve acceptance, and they often resist adult instruction along these lines.

It seems right to call the requirements for acceptability *needs*, but it is important to distinguish these needs (to be discussed in the next section) from the wants and interests that generate expressed needs. This distinction is important at every level from the personal to the global. Too often we assume that others feel the same needs we do, and we may feel both generous and righteous in working to meet needs that the recipients of our generosity do not endorse. On the global level, for example, Americans often act as though all people want individual freedom, but some clearly do not put a high value on such freedom. Similarly, we assume that everyone wants democracy, but there are cultures that prefer other forms of governance.

The negotiation of needs requires dialogue—talking and listening. At home and in schools, teenagers can profit from frequent dialogue on conventions, moral rules, and habits that regulate social life. Overly strict parents often treat all of these rules and conventions alike and engage in micromanaging their children's lives. In contrast, permissive parents may cause their children to live in confusion and uncertainty about what is expected of them in various situations. Parents should listen to expressed needs, show some understanding for the seemingly contradictory needs to belong and to rebel, and engage in dialogue with good humor.

There are at least two reasons for giving attention to expressed needs. One is political. Expressed needs are the forerunners of rights. When people express a want consistently and press authorities for its satisfaction, they may seek the power to convert their want to a right. This is often a long and difficult process, but once the right is established, people may regard it as something in place forever—God-given. Of course, people also make claims to rights that are frivolous—to walk on forbidden grass, to reject certain foods, to dress like a slob, to demand immediate service at commercial establishments, and to let their dogs run free. "I have my rights!" is a frequently heard exclamation in American life. The frivolous aside, the conversion of wants to rights requires not only persistence but the use of power. Since it is some form of power that is sought in a claim to rights, rights-seekers are at the mercy of those who already hold power. It is not surprising, then, that the process of moving from wants to rights is often accompanied by violence. For women—who are less likely than men to use violence—the process has often been marked by frustration and disappointment

A second reason for attending to expressed needs arises in parenting, teaching, and mentoring. Ruddick has named "fostering growth" as a paramount maternal interest, but again there are two ways of looking at this task.[10] In one way (a way I have identified with virtue-caring), the parent or teacher defines or presets the definition of growth and then provides the resources—and sometimes coercion—to move the child or student toward the definition of growth.

A second approach (identified with relational caring) is more open-ended. The adult in authority watches the younger person or beginner, listens to his interests, helps him to evaluate the goodness of his wants, and provides encouragement in the direction of the expressed needs. Notice that the authority figure is actively engaged in the process; she does not abdicate responsibility and say, in effect, "Do as you please." This second approach is a fundamental idea in progressive education and in progressive practice at the national and global levels. John Dewey put it this way: "There is, I think, no point in the philosophy of progressive education which is sounder than its emphasis upon the importance of the participation of the learner in the formation of the purposes which direct his activities in the learning process, just as there is no defect in traditional education greater than its failure to secure the active cooperation of the pupil in construction of the purposes involved in his studying."[11] This passage has to be read carefully. Dewey is not saying that students must be motivated by their teachers to do the work that is set out for them in a prespecified curriculum. He is saying that students must be involved in the *construction* of purposes for their learning. Such involvement requires the active engagement of both student and teacher. As a mathematics teacher, I should not insist that all students must like mathematics or even that they must do their best at it. I should find out what student purposes are served by their participation and how I might help them meet those purposes or, perhaps, revise them. It should be all right if some kids choose to do a merely adequate job at math (that's hard enough for many!) and save their creative energy for studies that interest them. I would not advise students to do their best in everything but, rather, to do their best at *something*. Good teachers and parents accept responsibility to assist the young in finding that something.

Working this way is not easy. We are continually challenged to increase our competence, because we are asked to assist in meeting needs for which we may not have expertise. It is not simply a matter of "learning with the kids," as some permissive educators advise. It is more a matter of continuing to learn, expanding our repertoires so that we are more likely to be prepared when a need is expressed. It is also a matter of admitting our ignorance on some topics and referring students to someone who can really help.

The approach just discussed is fundamental not only to progressive education but, even more basically, to care ethics. In listening and responding, we experience motivational displacement—a call to assist others in their projects. This is vital in all of ethical life but it is *central* in the so-called caring professions. In those professions, my project is defined in terms of helping others to meet their own legitimate needs.

At the time of this writing, the results of the 2007 SATs have just been published. Once again, despite increased participation in advanced school mathematics, girls score about thirty points lower than boys on the math SAT. They score a bit higher than boys, however, on the writing test. Almost certainly, there will be a renewed outcry to close the gender gap in math. Why? I mentioned this matter briefly in the introduction and again in the discussion of autonomy, but I want to reemphasize it now in the context of needs. We should not assume that all girls can and should do well in mathematics, especially not if such achievement requires them to relinquish or diminish their own interests. If girls who have aptitude and keen interest in mathematics are being discouraged in the subject, something should be done about their unfair treatment. If, however, the test scores are an indication that many girls do not see math as closely related to their purposes, perhaps we should ask more seriously what girls *are* interested in and evaluate their purposes generously.

Instead of reevaluating the interests and contributions of thoughtful girls and women, policymakers are now pushing hard for all students to take a second course in algebra despite the embarrassingly low scores now made by students taking the course voluntarily. Why do we insist on everyone studying and doing well in math when we ignore some of the most significant tasks of adult life? For example, although everyone seems to recognize that parenting is one of the most important and difficult jobs anyone tackles, schools rarely offer a course in parenting, and when they do, the course does not count toward college admission. You can bet that if men had been the primary caretakers of children for centuries, parenting would rank importantly in the curriculum.

I am keenly aware of arguments for reducing the gender gap in mathematics. Mathematics has long been defended as a gatekeeper to college. Students must take academic mathematics to gain admission, and they must test reasonably well to be accepted at selective institutions. Further, women still lag behind men in membership in the professions, in yearly earnings, in legislative and judicial bodies, and in managerial positions. Must they do better in mathematics to gain admission to full political and professional life? I do not see why. Many intelligent and highly creative people do not perform well in mathematics. A lack in this area does not seem to hold men back. It is time to recognize women's interests and contributions and reward them accordingly.

THE NEED FOR ACCEPTABILITY

Ruddick has suggested "shaping an acceptable child" as one of the three great maternal tasks. Obviously, a mother's view of what is acceptable will guide her efforts at shaping an acceptable child, and the cultures to which she belongs will in turn shape that view. In the context of needs, we can see that parents have a need to guide their children to acceptability, but how should we describe the child's corresponding need? In the previous section, I emphasized attending to expressed needs. Now it seems that we may be talking about imposing certain needs on children, students, and subordinates.

There are at least two contrasting ways to look at the task of shaping an acceptable child (student, worker, subordinate). In one way, familiar historically and still embraced in some traditional circles, the child is a sinful little devil who needs to be disciplined. The child's *need* is to be saved from himself—to save his soul. The other approach is to build on the child's need to be accepted and help him find a path to acceptability that we can endorse. This is not just playing with words. In the second approach, there is no assumption of original sinfulness nor any complete, detailed description of acceptability. There is a commitment to explore with the child the consequences of certain behaviors in light of the conventions of the culture to which the family belongs and to consider alternatives that avoid harm to self and others.

This way of thinking applies to global ethics as well. We should not assume that other cultures are necessarily deficient or sinful; nor should we assume that all societies should be converted to democracy or, as we once believed, to Christianity. We may try to move others to our way of thinking through dialogue, but in doing so we remain open to learning something from them as well. Nevertheless, we may insist that all nations *need* to abstain from torture and rape. Nations *need* to do this if they are to be acceptable in a circle of nations working together.

Reflecting on Maslow's hierarchy of needs, we might question various interpretations of *self-esteem*. Carol Gilligan found a "different voice"—a different path to mature morality—when she questioned Lawrence Kohlberg's stages of moral development; similarly, we may recommend different emphases when we consider the need for self-esteem. Self-esteem depends in part on acceptance by people we like and admire and in part on how well we live up to the ideals we set for ourselves. Schools have been ridiculed by conservative commentators for launching programs to promote self-esteem—"doing bad and feeling good." Critics are right to object to programs that fail to consider an appropriate base for self-esteem. However, we should not want kids' math scores to determine their self-worth. It is rational to assess one's math achievement realistically and face the fact that one is not doing very well. At the same time, it is not right to judge anyone's self-worth on the basis of such scores.

Instead of aiming directly at feeling good or insisting that self-esteem be earned by doing well at the tasks set for them by teachers and parents, children should be helped to seek acceptance in supportive groups, to pursue self-understanding, to grow in empathy, to appreciate relational autonomy, and acquire a host of social competencies. Empathic skills will open the door to supportive relationships, and self-understanding coupled with understanding of others should help young people to assess their own strengths accurately.

Education for self-understanding requires dialogue, and it also requires time to build caring relations. Good teachers show students that their worthiness as persons does not depend on their academic skills. It does not help to tell kids that they are doing well when they are not doing so. Too much praise—inaccurate and lavished on everyone by recipe—is confusing and dishonest. The best teachers and parents observe children carefully and take notice of the things at which they really do well. When a child helps another student or comforts an unhappy friend, a teacher may say, "That was a really nice thing you did." The compliment does not have to be announced to the whole world or supplemented with M&M's.

We want our young people to be good, that is, to be caring in their interactions and to comply with the conventions we consider supportive of moral life. But we should also want them to evaluate the conventions imposed on all of us and, with some humor, separate those that contribute to a life of caring from those that simply constrain us and make us picky. Adolescents are especially vulnerable to the demands of convention. On the one hand, they often rebel against adult conventions; on the other, they may slavishly follow the fads of their peers. Being different, belonging to a group that is different, can be very attractive to young people. Both belonging and being different seem central to adolescent life. Daniel Levitin writes that one of his professors at MIT "used to give out buttons that read 'Decadence Is Cozy.'"[12] Parents and teachers worry about this coziness and try to guide their children toward acceptable groups. It often helps for adults to discuss with them their attitudes toward the strictures of conventional adult life. What matters in life, and what is not very important? What do *you* think?

It has always been hard for some young people to achieve acceptance in groups that are themselves deemed acceptable. What makes a group attractive? What makes it acceptable and to whom? And how should we treat people who are not members of our in-group? It may be even harder today than it was a generation ago. Our schools are mired in a test mania that leads many students to suppose that doing well on tests should be their main objective in life. There is not much discussion about the good life and its many vital components.

To make things worse, many young people spend much of their time at computers, living virtual lives. A recent shocking report of online abuse appeared in the *New York Times Magazine*. Entitled "Mal*Web*olence," it told the story of "online trolls"—computer-savvy kids, mostly male—"who use the internet to harass, humiliate, and torment strangers."[13] Asked how he could justify the pain he causes, one "troll" responded that his victims were complicit in their own torture—they should "get over it" and ignore the nasty messages he sends. Earlier we discussed the importance of teaching children empathy (recall Hoffman's "inductions"), but I warned then that empathy without sympathy can be used for selfish, even hurtful, purposes. The "trolls" of "Mal*Web*olence," like many psychopaths, read their victims well; they know just how to hurt them.

The job of shaping acceptable young people is both personal and collective. There is a temptation to conclude that the "trolls" have been badly brought up, and one of the young men interviewed implies as much about his own upbringing. But the task is not only for parents; it belongs to the whole community and especially to teachers. Many policymakers, accepting life in bureaucracy's iron cage, would like to separate social functions from academic functions and assign to teachers the sole job of teaching a specific subject. In a caring community, we reject this emphatically.

As part of learning to think critically, kids need help in evaluating and resisting all sorts of peer pressure, including advertising. Instead of providing that important learning opportunity, many schools now deliberately expose students to advertising in exchange for free television sets and other electronic equipment. In schools, the educative principle should be to teach kids to think. Why are people so easily manipulated by advertising (and by propaganda)? How is psychology used to generate "needs" in toddlers, teenagers, men, women, and the elderly? Why do we not turn away in disgust from scenes of people brushing their teeth and gargling, svelte women draping themselves over shiny automobiles, men lamenting their loss of sexual readiness, people demonstrating the nonstickiness of their new denture holders, men rushing to the restroom to relieve themselves, and toddlers speaking in adult voices followed by vomiting for the camera? These ads apparently work. Why?[14] And why, when we cite critical thinking as an aim of education, do we not engage students in a critical examination of these practices?

In past generations, it was predictable that older people would tell the younger generation how much harder times once were—

grandparents walking miles in the snow to get to school and that sort of thing. As a child, I heard many of these stories and, in addition, had to listen to how grateful the poor Chinese would be to get the food I so easily rejected. We don't hear many such stories today—not honest ones at any rate. Although life is better today for some groups than it was a generation ago—for blacks, lesbians, gays, and the mentally handicapped—it is in some ways more difficult just to be a child. My grandchildren live in a much more difficult world—perhaps not as much physical work but surrounded by violent images and actual incidents, scurrilous language, indecent dress, broken homes, publicized corruption, and uncertainty. In addition to coping with all this, they are often faced with unrealistic expectations. To be respectable, one should take Advanced Placement courses, make high scores on the SAT, get into a good college, and enter a highly paid occupation. It isn't enough to be a good person and find a measure of happiness in everyday life.

I am not suggesting that we reorganize society to restore prudishness and stuffiness in social life. There is something positive to be said for the informality and honest talk characteristic of public exchanges today. But kids need help in evaluating the positive and negative features of this informality. They need to hear stories of mistakes we have made, not only stories of our heroism in overcoming obstacles. An ethic of care guides us to care most for our children, not so much about our image as parent or teacher.

In shaping acceptable children (workers, citizens), we want them to look outward to what others are going through. We help them to feel guilt and make restitution when they have hurt others. They *should* feel guilty for committing such acts.

However, we should avoid inducing shame. Shame, like pride, tends to turn inward and direct attention to oneself. It rarely helps to develop a life of care and concern for others.

I said at the beginning of this chapter that we would return to Braybrooke's concern about "embarrassing" needs such as addiction. They seem to be real needs; that is, they are expressed physically by those who have them. Braybrooke wonders how to fit them into a prioritized list of recognized needs. When we include the need for acceptability in our thinking, our problem is not how to fit embarrassing needs into some preconceived list but rather to ask whether meeting this need will make it impossible to meet an even more important need. If we simply satisfy the need for a drug, many other needs will go unmet. That realization suggests that the need for acceptability is extremely important and must get continuous attention. Our attitude toward addiction of any kind might be characterized as paternalistic (or maternalistic or parental), but that should not cause us to back away in horror. Sometimes we have to intervene when people are hurting themselves. This does not mean that we should condemn drug-taking or increase our ill-conceived and losing "war on drugs." If a person can take drugs, drink heavily, or gamble highly and still function in public and personal life, then we should not use coercion to change that person's habits. But when a person's ability to function is obviously impaired, we should place the need for acceptability (normal functioning in the relevant culture) first. Whatever we do to satisfy the embarrassing need should be carefully tailored to meet the more important need for acceptability. That, of course, is the intent of such programs as methadone treatment, but we have not developed them in a satisfactory way. In most cases, we just maintain people in a barely functioning stupor.

Much more attention must be given to the matter of "embarrassing" needs of every sort. A full discussion is beyond the scope of this book and beyond the expertise of the writer. However, the ethic of care with its emphasis on needs—basic, expressed, and assumed (or imposed)—gives us a way to approach this thorny nest of problems.

Where, then, does the ethic of care stand with respect to individual liberty? We find ourselves, I think, somewhere between the negative and positive conceptions of liberty described by Isaiah Berlin.[15] We cannot accept John Stuart Mill's dictum that we should not intervene in the life of an individual unless that individual is harming someone else; although we want each individual to exercise his or her limited autonomy, we feel a responsibility to intervene if he is harming himself. But neither can we accept a positive view of liberty that prescribes in great detail what a person should be and do. The emphasis on listening to and helping to meet expressed needs precludes our setting up preconceived ideals for all citizens or for any one person. The ethic of care recognizes our thorough interdependence.

The central message of this chapter is that dialogue—talking, listening, negotiating—is fundamental in identifying and interpreting needs. A second, vital message is that we should give respectful attention to expressed interests and use these to direct or assist people in meeting legitimate wants and needs. Even in the task of shaping acceptable people, we would be wise to start our work with expressed needs rather than those we might assume a priori. Wise diagnosis of expressed needs should help us in guiding people toward the needs for acceptability recognized by the culture in which we live.

So far I have acknowledged differences in male and female experience, and I have suggested that the contribution of female experience to moral development has not received the attention it deserves. Now we have to consider an element of male experience that should be changed—that may actually endanger the future of human life. What, if anything, can be done about the male tendency to engage in violence?

EIGHT

War and Violence

Perhaps the saddest evolutionary legacy still oppressing us is the male tendency toward aggression and violence. Many feminist thinkers have insisted that virtually all gender differences in temperament and behavior are products of culture and socialization, but evidence to the contrary has been accumulating. One need not be an essentialist to believe that many differences anchored in our evolutionary past are innate in the sense that they are heavily influenced by biology. When I reject essentialism, I reject the idea that female and male human beings were *created* with essential features by God and that these natures should not or cannot be changed. Deeply rooted evolutionary characteristics can and do change, but deliberate change is extremely difficult and may take a long, long time.

Care ethics aims to establish, maintain, extend, and enhance natural caring. It is thus clearly opposed to war and violence, but I have already admitted that it cannot advocate absolute pacifism. As parents, we will fight to defend the lives of our children, and in threatening situations, we will stand to protect those closest to us. Wishing to avoid violent conflict, our best hope is to mend breaches, meet needs, and educate for understanding across groups and between individuals. In this chapter, I will discuss the difficulties involved and why some popular remedies are hopeless.

JUST WAR THEORY

Just war theory has developed from the ideas of Augustine, who found it necessary to accommodate the demands of church and state. The basic idea is that a just war must have a just cause, be conducted by a rightly designated authority, entered as a last resort, and conducted with a reasonable likelihood of success, proportionality, and discrimination. There are two major sets of conditions—one describing just causes for entering war, the second setting out rules for the just prosecution of war. It has been argued that scrupulous adherence to the criteria for waging just war would make war rare.[1] However, we would be hard put to find a historical example of war conducted justly.

It is, perhaps, easier to claim a just cause than to prosecute a war justly. World War II, for example, is often referred to as the "good war," and as time erodes memories, it becomes even better. Daniel Levitin, for example, describes his understanding of World War II when he was seven years old: "A tyrant was trying to kill all the Jews; we were Jewish, and some countries came to our aid. That war made sense."[2] It would indeed have made sense if the world had gone to war to rescue the Jews, but it did not, even though some adults continue to believe that it did. The United States fought because it was attacked (creating a just cause), and it did ultimately fight against those who had instigated a horrible genocide, but the slaughter of innocents was not the cause of our entry into war. If we probe deeply into the causes of the war between the United States and Japan, the case becomes muddier. Claiming a just cause implies that someone else did something unjust, and it isn't always clear who has the better argument.

But, even if we could pin down and agree on just causes, the theory falls apart when we try to apply it to just prosecution or conduct. In defending just war theory, Michael Walzer writes of the soldier's proper position: "They can try to kill me, and I can try to kill them. But it is wrong to cut the throats of their wounded or to shoot them down when they are trying to sur-render. These judgments are clear enough, I think, and they suggest that war is still, somehow, a rule-governed activity, a world of permissions and probabilities—a moral world, there-fore, in the midst of hell."[3] We could challenge the notion that a rule-governed activity, one of permissions and probabilities, is necessarily a moral world. Indeed, I've argued throughout this book that morality is not best defined in terms of rules. But in any case, what good are rules if they cannot be followed? The "good war" was loaded with horrible cases of breaking the rules on all sides.[4] If there is something in human nature—male nature especially—that makes it impossible or at least highly unlikely that the stated rules will be obeyed, then we would do well to find a way to eliminate the game. If human nature makes it impossible to do this, then we must begin to consider the pos-sibility of changing human nature, and it may actually be a little easier to move men toward modified pacifism than to shape them to obey the rules of war. Further, we should seriously ask what kind of people we have become when we are morally satisfied by setting rules under which it is appropriate to kill one another.

Walzer makes a sharp separation between what he calls the *war convention* and the conduct of war. This too may be a mistake. He writes: "We cannot get at the substance of this convention by studying combat behavior, any more than we can understand the norms of friendship by studying the way friends actually treat one another."[5] Walzer is discriminating between prescriptive and descriptive rules or principles. This is in itself right, but we should note that prescriptive rules or norms are derived from an evaluation of best practices. Then over time the norms are modified, sometimes idealistically away from practice and other times more realistically in the light of practice. Care ethics advises continuous observation of real conditions and behavior, the recognition of turning points, and close adherence to reality without simply accepting the "way things are." The conventions of war are of little use if they are regularly contravened. Moreover, it is hard to argue that an immoral activity can be conducted morally.

Is it feasible to suppose that soldiers can be taught and monitored to obey internationally agreed-upon rules of war? Walzer concludes: "The transformation of war into a political struggle has as its prior condition the restraint of war as a military struggle. If we are to aim at the transformation, as we should, we must begin by insisting upon the rules of war by holding soldiers rigidly to the norms they set. The restraint of war is the beginning of peace."[6] It won't work. First, it isn't only soldiers who must be held to a rigid standard. It is more important that national leaders espouse and uphold the standards, and although they give lip service to the standards of proportionality and discrimination, they often intentionally override them. The war convention has long forbidden intentional attacks on civilians, and yet civilian casualties have risen enormously in wars over the last century. Indeed, in World War II, the Korean War, Vietnam, and Iraq, civilian casualties far outnumbered those of the military. Sometimes civilian casualties are accidental—"collateral damage"—but often leaders attempt to justify casualties that could have been avoided by using ethical principles outside the war convention.

Walzer discusses (without defending it) the American decision to drop atomic bombs on Nagasaki and Hiroshima. Under the war convention interpreted rigorously, such a move should not even have been considered. But President Truman and his staff brushed aside the war convention in favor of Utilitarian calculations. Killing thousands of innocent civilians was justified as a means of saving even more lives, and the figures are still debated. Did it save lives, or did it not? We are reminded of the dilemma in which we are asked to consider pushing a heavy man onto the trolley track in order to save five other lives. We should not even consider it. We have a prior commitment to those who stand beside us or right before us at our mercy. I am not a Kantian, but on this, Kant and care theorists agree.

Walzer argues strongly for the *rights* of civilians caught in war. He recognizes that in modern warfare—aerial bombing in particular—civilian casualties cannot be entirely avoided. They should, he says, be minimized. Civilians have *rights*. But rights are demanded and granted by human beings under certain conditions, and they are ignored or retracted under other conditions.

Care theorists do not despise or belittle rights. But we want to stay in touch with needs, wants, and desires: the need for food and safety, the desire to see one's children grow up and prosper, the universal want to avoid pain and deprivation. Women opposed to war have long insisted on this emphasis, and in World War I, women's organizations reached across enemy lines to support women and children.[7] For these women and for many peace activists, it was not a matter of rights but a recognition of universal human needs and a deep respect for maternal interests.

A second reason why the idea of learning to fight by the rules won't work is that some men who enter the military are actually attracted to war and violence. For them, military action provides sanction for horrible acts otherwise forbidden. It provides first, and most innocently, the "right" to use language condemned by polite society and the *New York Times*. Listen to Anthony Swofford introducing his experience with the marines in the Gulf War. He notes that while many thoughtful citizens are persuaded against war by antiwar films, his marine comrades react very differently: They "are excited by them, because the magic brutality of the films celebrates the terrible and despicable beauty of their fighting skills. Fight, rape, war, pillage, burn. Filmic images of death and carnage are pornography for the military man; with film you are stroking his cock, tickling his balls with the pink feather of history, getting him ready for his real First Fuck. It doesn't matter how many Mr. and Mrs. Johnsons are antiwar—the actual killers who know how to use the weapons are not."[8] One of the great attractions of military life for many young men seems to be the freedom to use language forbidden to them as children. Why does the word *fuck* appear in almost every utterance of men in the field? A whole chapter in Dexter Filkins's powerful account of the Iraq war is entitled "Fuck Us."[9] Perhaps it conveys the truth that the men are about a filthy business, and the language fits the activity. Perhaps there is no time for reflection on better modes of expression. Perhaps it is a sign that "real men" defy the mores of polite society. Perhaps it is a way of avoiding gentle feelings, of breaking down in tears over what they see and do.

One can argue, rightly, that most men who enter the military are not like those described by Swofford. More young people join the military to experience some adventure and find meaning in life. War connotes risk, bravery, recognition, and belonging. Chris Hedges writes: "The enduring attraction of war is this: Even with its destruction and carnage, it gives us what we all long for in life. It gives us purpose, meaning, a reason for living."[10] J.G. Gray also speaks of "the enduring appeals of battle" and "the delight of destruction."[11] This enduring appeal is not limited to those described by Swofford who actually revel in cruelty. Gray recounts incident after incident in which normally rational men are driven to a "monstrous desire for annihilation." Drawing on both his research and his personal experience in World War II, he writes: "Most men would never admit that they enjoy killing, and there are a great many who do not. On the other hand, thousands of youths who never suspected the presence of such an impulse in themselves have learned in military life the mad excitement of destroying."[12] The impulse noted by Gray is almost certainly a product of evolution—a widespread tendency in both male humans and chimpanzees.[13] I'll say more about this in the next section on gender and violence. But if we take this evolutionary legacy seriously, we should conclude that societies must find a way to limit violence either as Walzer suggests—by enforcing the rules of war— or by finding a feasible way to eliminate war.

William James, shocked and horrified by what he called history's bath of blood, wrote that we had to find a "moral equivalent" of war (an unfortunate choice of words): "What we now need to discover in the social realm is the moral equivalent of war: something heroic that will speak to men as universally as war does, and yet will be compatible with their spiritual selves as war has proved itself to be incompatible."[14] But is war incompatible with spiritual selves as defined by religion? In today's world, war and religion are still often mutually supportive. Terrorists think they are doing God's work, and several American leaders believe we are doing God's work in Iraq. To find a humane solution, we might seriously consider abandoning both war and religion. Remember, it is a warlike, vengeful *male* god who is worshipped in the three great monotheisms.

I want to mention one more reason that Walzer's plan won't work. The stakes are too high today. Since World War II, when it became possible to kill hundreds of thousands of human beings with one bomb, the idea of a plausible war convention has faded. Pushed hard enough, nations will use the resources over which they have command.

Only "police actions" can reasonably be guided by an agreed-upon convention. Applied to such actions—and there should probably be more of them to prevent genocide and other grave injustices—Walzer's recommendations are highly useful. But the use of force to prevent violence must be meticulously monitored. Police too can give way to unauthorized violence. However, it should be possible for a police action to be conducted by rules because the enormous baggage of patriotism—fighting for one's flag, defending one's country, upholding the honor of past warriors—is removed.

Recognizing the deeply ingrained tendency to male violence, it is surprising that few commentators say anything about education. Walzer says virtually nothing on it. Gray writes eloquently on the human mind and how it must change, but he says nothing about the role of education in effecting the change.

VIOLENCE AND GENDER

When we are faced with a huge problem, it usually helps to understand its parts and admit its difficulties. Too many people—many thoughtful feminists among them—approach violence as though it is entirely a product of culture, primarily of socialization. That culture and socialization have a powerful influence there is no question. The work of psychiatrists such as James Gilligan documents the effects of abusive childrearing on adult behavior, but when criminal violence follows childhood abuse, it is almost always in *males*. Females rarely become vicious killers as a consequence of childhood abuse.

Men start and fight wars. Despite exotic stories of female warriors, the historical record confirms that wars have been men's wars. Within a society, males commit the overwhelming majority of violent crimes, and this is true worldwide. "Male criminals specialize in violent crime. In the U.S., for example, a man is about nine times as likely as a woman to commit murder, seventy-eight times as likely to commit forcible rape, ten times as likely to commit armed robbery, and almost six and a half times as likely to commit aggravated assault."[15] Men are naturally more aggressive and violence-prone than women. Why is it so hard to admit this, and what difference might it make if we did admit it? Feminists have, understandably, been reluctant to consider that male violence and domination are artifacts of evolution. Such an admission seems to saddle us with "essentialism," the doctrine that males and females have been created with essentially different natures. The problem with this doctrine, as I noted earlier, is twofold. The first comes with the word *created*. If we believe that a male god created males and females with the traits so widely observed, then it follows that we ought not to mess around with them. We must accept them and concentrate on what is good in them. Feminists are rightly outraged by this. What woman (besides, perhaps, Phyllis Schlafly) wants to celebrate the good features of women's subordination? The mistake is in supposing that the natures so described were *created*. If we interpret them as products of biological evolution, aggravated by cultural evolution, we can see that they will be very difficult to change, but they *can* be changed.

Hard as it is for many women, we must consider the possibility that there is no Father-Creator-God, and we are thus free to pursue other spiritual paths or to give up on the spiritual. One can still retain the most beautiful of religious recommendations, modified by reality: Love one another (as the God we wished for might have loved us). This injunction flows more naturally as an extension of maternal instinct than from the hand of the warlike male god.

The second part of the problem is that essentialism has always worked in favor of males and against women. Attributes and traits associated with males have been elevated; those associated with women have been denigrated or smiled upon benignly (the "isn't she cute?" response). This tendency can be opposed through education broadly conceived, but it first has to be faced and discussed intelligently.

Earlier I discussed the possibility that females (in general) may have less interest in objects and mathematics than males (in general). This does not mean that there are no women who have exceptional talent in mathematics, nor does it mean that there are not a good many women who have sufficient talent to enter careers in mathematics. It should remind us, however, that many young women will have interests that are not mathematical, and we should not despise or belittle those interests. Why are mathematical interests so highly valued? Mathematical achievement has made incalculable contributions to human progress. Yes, but so have the virtues and skills associated with maternal instinct. Why is the first set so much more highly valued than the second? The answer seems to be that the first set has been the province of men. It is understandable, then, that feminists want to insist that there are no innate cognitive differences between males and females. It seems more promising to say that it is a matter of cultural persuasion and discriminatory socialization.

Do we want to argue that male aggression and violence are solely products of culture and socialization? Certainly they are strongly supported by features of our culture and socialization patterns, but the tendencies themselves are almost surely products of evolution, and it will take extraordinary educational efforts to modify them.[16] Put boldly, the problem is that people put a high value on the aggressive tendencies of males. We even judge a woman's fitness for leadership by her toughness—how nearly her manner approximates the assertiveness of males. It is nearly acceptable for a woman to act like a man, but it is not acceptable for a man to act like a woman.

In written history, whole societies may be omitted or given a much-reduced role if they have not displayed male aggressiveness. Clark Wissler, for example, comments on the Hopi: "Probably because of their long passive resistance to white influence, they have few heroes known to history. Wars seem necessary to reveal greatness. Had the Pueblo terrorized the settlements, massacred women and children, left a trail of blood and destruction behind them, they would hold a high place in history, as we know it."[17] The high value put upon war and aggressiveness is just one, very prominent, feature of the larger picture. Everything male sweeps away those matters of greatest interest to the female. Anthropologists note the female understanding and support of relatedness, but they rush on to an investigation of alpha males, status hierarchies, and military-like bonding. Even in Daniel Levitin's description of singing among the Mekranoti Indians, the women are somehow lost. Levitin writes intriguingly: "One of the most remarkable things about the Mekranoti is the amount of time they spend singing—women sing for one or two hours every day and men sing for two hours or more each night."[18] It turns out that the men sing—in a "masculine roar"—to stay awake and fend off possible attackers. Levitin elaborates on this custom by describing the widespread Native American use of song in preparing for battle. He notes also the fearsome use of drums and bagpipes by the Scots and the terror-inducing war cries recorded in the Old Testament. I read this section eagerly—waiting, looking forward to hearing why the Mekranoti women sing and what they sing about—but there was nothing on this. The women's songs were lost, apparently irrelevant. When I asked him about this by e-mail, he said that the women's songs were "typical"; but it seems to me, as a woman, that the men's songs were typical. I'd still like to know what the women sang about. Were they singing about love? Children? Cooking? Harvesting? Later in the book, one page is devoted to lullabies, the first songs human beings hear, but that discussion ends inconclusively with a debate between two theorists (both male): Is the lullaby intended to calm the infant or the singer? Ho, hum. It seems more important to get back to alienation, sexual love, resistance, violence, and heroism.

In fairness, Levitin celebrates song that expresses the "love of our existence that is the highest love of all, the love of humanity with all our flaws, all our destructiveness, all our petty fears, gossip, and rivalries."[19] But there is still something missing. Destructiveness, fear, rivalry. What of tenderness, protectiveness, nurturance?

In contrast to the male celebration of violence and military heroics, maternal thinking—free to express itself—despises war and violence. But maternal thinking is not always free to express itself. Women are often complicit not only in their own subordination but in the exploitation of their children. Gold Star mothers are torn between pride and sorrow. In most cases, they should be angry, even furious. But the social pressure to be proud of their sons and to be patriotic overwhelms their justified sense of injury. They are assured in war after war that their husbands and sons "did not die in vain." A clear-eyed assessment would tell them that, in most cases, their loved ones—like all victims of war—did indeed die in vain.

It cannot be claimed, then, that women always oppose war. In fact, they have sometimes been vociferously enthusiastic about it.[20] One obvious reason for this collaboration is the desire to belong and to demonstrate patriotism. But many women have been articulate in peace movements, and they have often adopted the stance of World Citizen despite the pressure of caring for those nearest to them. Writing of the period during and after World War I, Jean Bethke Elshtain notes: "Women peace campaigners promulgated internationalism, as a worldwide concatenation of peace-loving peoples, especially women, to bring into being the conditions for permanent peace. After the war, women were influential in pressing for the Kellogg-Briand Pact (1928) declaring war illegal and were on the National Committee on the Cause and Cure of War, which collected ten million signatures on a disarmament petition in 1932."[21] It is not my purpose here to discuss the history of women's involvement in peace movements, but it is both thrilling and disheartening to know that women tried eighty years ago to make war illegal.[22] At this point I simply want to emphasize the need to recognize and address the evolutionary differences between women and men and what this recognition might mean for ethics and moral efforts to achieve peace.

We might pay particular attention to the difference in language used by male and female peace activists. The men are often theorists, working out and defending positions on war and peace; refusing to commit violence, they frequently invite its use against themselves. They become willing (heroic) victims of the violence they deplore. Women are more likely to be directly involved in meeting needs, including the need to protect themselves and their children. Their language refers to their own hands-on interactions with real people whose needs they are trying to meet, not to events through which they may be acclaimed as heroes.[23]

There is also a special sensitivity among women to bodies and flesh. Pearl Buck wrote of her parents' contrasting views. Her father, a Presbyterian minister, was concerned with souls, almost brushing off the physical deaths of children, saying, "Doubtless it was the Lord's will and the child is safe in heaven." But her mother, Carie, agonized over both dead children and bereaved mothers. Buck wrote: "Once I heard someone say of another's dead child, 'The body is nothing now, when the soul is gone.' But Carie said simply, 'Is the body nothing? I loved my children's bodies. I could never bear to see them laid into earth. I made their bodies and cared for them and washed them and clothed them and tended them. They were precious bodies.'"[24] A similar sentiment was expressed by Olive Schreiner: "No woman who is a woman says of a human body, 'it is nothing.' . . . On this one point almost alone, the knowledge of woman, simply as woman, is superior to that of man; she knows the history of human flesh; she knows its cost; he does not."[25]

Sara Ruddick gives many more examples of this special sensitivity, mentioning especially the celebratory nature of giving birth. But then she adds: "Oddly, there is barely an echo of this celebratory aspect in philosophical writing. Although we are a species that knows its own natality, in philosophical texts we are 'thrown' into the universe somehow, appearing at the earliest when we can talk and read."[26] Ruddick provides an interesting analysis of this philosophical neglect, but at bottom it is all part of the long-standing, pervasive denigration of women. Birth is painful and messy; therefore, man needs somehow to be "born again" without a physical body created through the pain, nurturance, and love of a physical mother. With tremendous effort, he leaves behind the physical and ascends to the world of reason and abstraction. Having done so, he supposes that everything associated with his physical origin is somehow inferior, and morality itself must be a product of abstract reasoning—not an achievement of concrete reasoning carefully connected to the origin of life in maternal instinct. Care ethics requires reasoning as it moves from maternal instinct to natural caring and ethical caring. At the level of ethical caring, it turns back appreciatively to natural caring and employs both reason and feeling to find ways to restore and extend the conditions that support natural caring.

WHAT SUPPORTS WAR?

All sorts of social/political explanations of war fill our libraries and textbooks. Social studies schoolbooks are organized around war (e.g., *American History from the Revolution to the Civil War*), and children are often required to learn the remote and immediate causes of particular wars. A main cause—the evolutionary male trait favoring aggression—remains unmentioned.

Virginia Woolf once said that the cause of war is "manliness and manliness breeds womanliness—both so hateful."[27] However, if we believe that manliness and womanliness are both rooted in evolutionary tendencies, we know that they will be hard to change. We have no choice but to work on the social/cultural customs that support the evolutionary base, but powerful forces work against us. Even men who are strongly opposed to war inadvertently support it by accepting traditional views of masculinity. William James, in his moving plea for a "moral equivalent of war," asked "whether this wholesale organization of irrationality and crime [war] be our only bulwark against effeminacy."[28] The loss of manliness, it seems, would be a terrible price to pay for the rejection of war. The idea, evidently, is to get rid of war but maintain the manliness that makes man superior to woman. Can we imagine complimenting a man by saying that he is as loving and nurturing as a woman? Yet we regard it as a compliment when we say of a female politician that she is as tough as a man.

One might reply to this line of argument that reference to the masculine-feminine difference does not connote superior-inferior but, rather, acknowledgment of types. When a man is *masculine*, he is true to type. When a woman is *feminine*, she is true to type. I have already accepted major differences traceable to our evolutionary heritage; so what is the complaint? The problem is this: If we make a long list of "masculine" and "feminine" attributes or traits, we will find that those in the masculine list are valued above those in the feminine list. It is becoming more and more acceptable for a woman to adopt traits on the masculine list, but a man who behaves like a woman is mocked and scorned.

Our patterns of socialization are powerful in maintaining masculine and feminine types. Michael Kimmel remarks, "The rules of masculinity and femininity are strictly enforced. And difference equals power."[29] We could argue that the gender traits stemming from evolution should be reevaluated on the basis of fairness; if they are accepted as traditionally described, women are treated unfairly. Therefore, to be fair (some argue) we should recognize that women can exhibit the same admirable qualities as men—hold public office, exhibit courage, excel in mathematics, serve in the armed forces, engage in contact sports. But we could argue more powerfully that the much-admired masculine type is too dangerous for the health of humanity and must be transformed. The only hope for influencing our own evolution is to exercise some control over the conditions that support the male tendency to aggression.

Consider the widespread valorization of patriotism. To be a patriot, one must love, support, and defend one's country. In the context of our earlier discussion of World Citizen and Charity Begins at Home, patriotism often tips the scale toward the latter, and we have already noted current work on altruism that documents the biological tendency of humans to protect and support those closest to us. Candidates for national office almost universally promise to do what is best for their own country. In the United States, this attitude is pushed to the extreme of striving to be (or claiming to be) number one, first in all things highly valued. If others also benefit, well and good, but our own nation comes first. This attitude is far more likely than care ethics to endorse Charity Begins at Home. Care ethics insists on caring for those with whom we have direct contact, but—in caring about—it also seeks to expand the circles of care and to establish new ones. It is critical of characters like Dickens's Mrs. Jellyby who "care for" people far away and neglect their own children, but in trying to extend the circles of care, it is likely to align itself with Thomas Paine, Virginia Woolf, and Jane Addams in seeking community beyond our national boundaries. And we know how a patriotic public reacted to those folks. Thus we see a paradox: care ethics recognizes that we humans are so constituted that we will stand up for those closest to us even when they are wrong, but it does not approve of this tendency. Male-defined patriotism makes a moral good of the tendency.

How do we inspire patriotism? Patriotism is almost always inspired uncritically. Young children are strongly invited to recite the Pledge of Allegiance at an age when many of them cannot even pronounce the words correctly. From an educational point of view, it would be better to omit this group exercise until sometime in middle school and then introduce it with a critical history. How and by whom was it commissioned? Who wrote it? What was its purpose? Who objected and what was the official response to objections? When were the words "under God" added? Why?

This discussion might be followed by a brief exploration of how people in other countries express devotion to their countries. The aim is not to undermine love of one's country but to gain an appreciation for the way others feel about their countries and begin to explore what this love means for human solidarity and moral life. We should want kids to care about those they do not meet directly and begin to understand that people in other nations care for one another as we care for our own. This is a simple but profound lesson. Could we muster support for it?

It has been suggested periodically that we abandon the "Star-Spangled Banner" as our national anthem and replace it with "America, the Beautiful." The latter contains no warlike references to bombs and rockets and includes a beautiful plea for brotherhood. It is also easier to sing! But even this mild, constructive suggestion has been vigorously rejected. Inducted early and uncritically to the rituals that support patriotism, many people are unwilling to turn an analytic eye on them. Again, as students sing and read the words of "America, the Beautiful," they can explore how people in other lands feel about their countries and consider the question, Can brotherhood be extended beyond "sea to shining sea" to the whole world? Readers should also note that "America, the Beautiful" was written by a woman, Katherine Lee Bates. Why is it so hard to change the national anthem? When the Pledge of Allegiance was written (1892), the "Star-Spangled Banner" had not yet been designated as our national anthem. Although it was written in 1814, it was not officially named our national anthem until 1931.

The newly awakened interest in global warming, species' extinction, and overconsumption of resources should provide considerable motivation to move us beyond narrow national interests, but even with respect to such global problems, much of the required work will be inspired by concerns for "our country." It may not be concern for others that moves us so much as concern about what the condition of others means for our own welfare.

Heroic stories, parades, uniforms, flags, and holiday celebrations all support patriotism and, unfortunately, war. How many of us would willingly give up the celebration of July Fourth? On the New Jersey shore, where I live now, the day is magnificent—flags flying, parades, kids scurrying for candy scattered by parade participants, old cars, clowns, fife-and-drum corps, picnics, fireworks, and Sousa from morning until night. I admit to loving it. But I think about what the spectacle supports, and I want my children and grandchildren to think about it. Why do we so often celebrate with guns and mock battles? Why is it so often claimed that unless we fight again (or continue to fight), those who died in past wars will have "died in vain"? And think, think. Is it not possible that many of them (and hosts of innocents with them) did in fact die in vain? Perhaps our objective should be to eliminate war so that no one need die to ensure that those already dead did not die in vain.

The persistent social admiration of masculinity itself supports war. In a chapter that should be required reading for all young men, Michael Kimmel describes the status of masculinity and the behaviors it encourages.[30] It is a frightening description of sexual entitlement, aggression directed at women, and power relations used for domination. My only disagreement with Kimmel is sparked by this statement: "Sexual beings are made, not born."[31] Sociology went through a stage when this view was popular, but the evolutionary sciences are now leading us to question the assumption that masculinity is entirely a matter of culture and socialization. Unquestionably, many cultural practices encourage male aggression, and it will not be easy to rework (or abandon) the traditional view of masculinity. But some practices might provide harmless outlets for aggression. Which ones? That is a question we are still struggling to answer.

Religion must also be named as a strong supporter of war and violence. At the beginning of the twenty-first century, we hear exhortations from Islamist extremists to fight and kill unbelievers, threats of death to those who oppose Jewish extremists in Israel, battles between Hindus and Moslems in India, and a destructive, preemptive war led by a Christian president who feels guided by God to spread democracy over the Middle East. Most religious traditions say that they stand for peace even though they have violent histories. Even Buddhism to which many young people today are attracted because of its claims to peace is not exempt from a bloody history. Christopher Queen writes: "One should . . . note at the outset that violence has not been unknown in Buddhist societies. Wars have been fought to preserve Buddhist teachings and institutions, and Buddhist meditation and monastic discipline have been adapted to train armies to defend national interests and to conquer neighboring peoples."[32]

So much has been written on the connection between war and religion that I will not belabor the topic. Sometimes religion has been the main instigator of war and violence. Often, while deploring a war and leading prayers for peace, its institutions have stood solidly with the nation-states in which they were located. Religion adds the authority of God to that of a nation's leaders and provides justification for acts that would otherwise be criminal. Consider the contradictions that appear in the following paragraph from Jared Diamond, writing of the benefits organized religion brings to societies: "First, shared ideology or religion helps solve the problem of how unrelated individuals are to live together without killing each other—by providing them with a bond not based on kinship. Second, it gives people a motive, other than genetic self-interest, for sacrificing their lives on behalf of others. At the cost of a few society members who die in battle as soldiers, the whole society becomes much more effective at conquering other societies or resisting attacks."[33] One hardly needs to add comment to this. Religion forbids men to kill one another within a tribe or nation so that they may better and more righteously kill members of other societies. If religions are to prove their commitment to peace, they will have to go through their histories and texts with an attitude of penitence and eliminate or soundly reject all of the doctrines, paragraphs, sentences, and stories that support violence. Having done that, there may not be much left to command belief.

Along with religion, education (schooling) in most parts of the world supports war. It promotes patriotism, obedience to authority, respect for religion, and—in its choice of curricula—masculinity. It usually purports to advocate and teach critical thinking, but it often forbids teachers and students to raise and discuss critical issues. Still, it may be our only mechanism for desirable change, and we'll explore that possibility next.

WHAT EDUCATION MIGHT DO

Education is supposed to enrich the lives of students and prepare them for the future. For a significant number of students, the period after high school will include service in the military, but our schools fail dismally in preparing students for what might happen to them if they engage in combat. Students know, of course, that they might lose life or limb in combat, but no one warns them that they might lose their moral identity—that they might *do* things that will traumatize them for years, perhaps for a lifetime. They are rarely asked to think reflectively on stories such as that of Karl, the dying Nazi, discussed in chapter 5. What does it mean to lose one's moral identity, and how does it happen?

There is powerful evidence from both historical and psychiatric accounts that ordinary men—men who would never harm their neighbors or commit crimes—do terrible things in war. Imagine how a young soldier feels when he throws a grenade into a house where a sniper is thought to be hiding. After the grenade explodes, he enters the house to find the bodies and body parts of children and their mother. A normally decent young man will find this hard to shrug off as simply an accident of war. Many veterans of the Iraq war suffer mental illness because they can't come to terms with the awful things they did as warriors. A significant number of Vietnam veterans still suffer such agonies.

Consider how much worse the trauma will be when horrendous results are the product of loss of control, not accident. When men go mad with fear, hatred, and fury, they commit acts for which they suffer self-hatred. Jonathan Shay, a psychiatrist who has been treating Vietnam veterans for years, has compared the loss of control exhibited in Vietnam to that of Achilles raging on the battlefield after the death of Patroclus. We have learned little in twenty-seven centuries about the male psyche in war, or if we have learned quite a lot, we have also learned to ignore it. The subtitle of Shay's book, "Combat Trauma and the Undoing of Character," is revealing.[34] The warrior who loses control and kills indiscriminately is sometimes regarded as a hero if his victims are enemy soldiers but as a monster if they are civilians. He has lost his character, his moral bearings. One veteran in Shay's therapy group described his "good luck" in being assigned to a battle site where there were no civilians. His unit did horrible things to enemy soldiers (hence his need for therapy), but he expressed horror at the thought that his fellow soldiers could do such things to women and children. However, his companion in dialogue said: "Well, at first, I mean when I just come there, I couldn't believe what I was seeing. I couldn't believe Americans could do things like that to another human being . . . but then I *became* that. We went through villages and killed everything. I mean *everything*, and that was all right with me."[35] Obviously, it was not permanently "all right" with him. He had lost his moral identity and might never recover it, but the loss troubled him. The issue of moral luck arises here. Some of us, even some combat soldiers, are never put in the situation that triggered the loss of moral identity in the veteran quoted above. Shay remarks: "Our culture has raised us to believe that good character stands reliably between the good person and the possibility of horrible acts."[36]

Character is not enough, however, and our young people deserve to know this before they put themselves at risk for the loss of moral identity. Michael Walzer apparently also believes that good character supplemented by the meticulous control of commanders in the field can keep men firmly committed to the accepted rules of war. History and present military activities tell us a different, chilling story. Not only is a loss of moral control likely in any battle situation; its likelihood grows when the enemy is noticeably different from us in color, physical conformation, or culture. Students study "about" World War II, for example, and learn that it is often referred to as the "good war," but they are not informed about the vicious fury of wild American troops engaged against Japanese enemies. Peter Schrijvers gives us vivid accounts of American atrocities committed against enemy soldiers, hapless civilians, and even the environment in the Pacific war, and he comments, "Rampant rage failed to be extinguished even by orders from above."[37] When men are thrown into war, their moral luck often runs out and with it their moral identity.

I am not arguing that exposing high school students to these bone-chilling facts will prevent the acts associated with the loss of moral resources. Care theory acknowledges human nature and its limits. We cannot depend on universal moral heroism. If we could, it would not be heroism but ordinary behavior. The only way to prevent violent atrocities is to prevent the conditions that produce them. We have to create an environment in which *it is possible* to be good. However, it is part of our responsibility as educators to prepare our students well for their possible futures, even those futures we deplore. Young men who go directly to college from high school have a better chance to read and consider the material I've been discussing here and are likely to make more well-informed choices, but even they—aware of what may happen to them—will not be immune to violence in conditions that support horrors. High school kids going directly into the military simply do not know what they are getting into. The situation may be even worse for the 30 percent of new army recruits who have not graduated from high school.

Schools could also make a valuable contribution to the cultural evolution of masculinity by encouraging boys to consider careers in the caring professions. Including more males in social work, nursing, early childhood education, and elementary school teaching would gradually reduce the fear of effeminacy and the notion that caregiving is women's work. Preparing for this work should also promote a more genuine understanding of caring and the ethic of care. Caregiving, as noted earlier, may act as an incubator for caring. We often learn to care and to become caring as a moral way of life through the work of caregiving.

The apprenticeships of future caregivers should be well supervised. It is worth repeating that occupational caregiving is not synonymous with caring. There are "caregivers" who are cold and uncaring, and candidates who exhibit such tendencies should be weeded out. The school's part in increasing the number of males in caring occupations would largely start in counseling offices and clubs such as future teachers, future nurses, and the like. Working together, with the support of administration, they could assemble a group of mentors from the caregiving professions and establish an apprentice program. Such a program would be a valuable community service as well as a learning experience for the students who participate.

There should be further benefits to recruiting more males to the caregiving professions. The presence of males might well drive up the salaries of all caregivers—a welcome effect. But opening these occupations to males would also introduce new opportunities for boys, and that is an important result at a time when so many boys are disaffected by schooling and discouraged about their own futures. Many boys might welcome the chance to show that "real men" can be as tender and compassionate as a woman.

The most difficult challenge for the schools would be to analyze and reconstruct the whole curriculum to make it inclusive of female experience. I am not optimistic about accomplishing this. Consider the present direction of schooling. Policymakers in many states have decided that all children now must study algebra, and in some states (New Jersey is one), a second year of algebra will be required. This is unrealistic to the point of the ridiculous. Even the youngsters who take the second year of algebra voluntarily have a hard time with it, and standardized tests on the material have produced deplorable results. Moreover, we graduate from high school huge numbers of students who have no sense of everyday financial management and are likely to add to the horde of adults who cannot pay the bills on their credit cards or mortgages. In a pessimistic mood, we might suspect that schools are complicit in maintaining an ignorant society of free-spenders.

If women had been involved from the start in the development of school curriculum, courses on parenting, home management, child development, life's stages, meal planning, gardening, healthy lifestyles, and food handling would certainly have appeared in our schools.[38] All of these topics have a rich intellectual dimension as well as immediate practical importance.

Only a wild-eyed dreamer would suppose that such a revolutionary change in the curriculum could actually take place. The entire structure of schooling stands as an immovable barrier. Higher education calls the tune—establishing "rigorous" requirements for admission that in turn dictate the school curriculum. Pseudocourses in algebra and science (our schools are rife with them now) are more acceptable to colleges (the transcripts *say* algebra and geometry, after all) than a course in parenting— even though the latter might be rich in history, literature, cultural geography, biography, and critical thinking. The school curriculum is a powerful arm of centuries-long male hegemony. To their credit, male-based theories of justice have now insisted that females may share in the established forms of education. It is acceptable for women to become more like men. It should also be acceptable for men to become more like women. Our future as a species may depend on it.

In this chapter, we've looked at the evolutionary legacy of male aggressiveness. We've considered the possibility of conducting war justly, the social customs and practices that support war, and what schools might do to transform the traditional pattern of masculinity. In the next and final chapter, we will consider what might be gained from a convergence of masculine and feminine views on moral life and whether such a convergence is practically possible.

NINE

Convergence

Throughout this book, I've been exploring caring, morality, and the development of an ethic of care from their roots in maternal instinct. However, I have not claimed that there is only one evolutionary path to morality. Morality has long been defined by some in terms of God's word. Obviously, there is also the well-known and well-trodden path from original self-interest. Men have learned that fairness and cooperation often increase their own chances of survival, and that train of thought has led to elaborate (and sometimes competing) systems of moral thought. I have argued that these traditional systems frequently go too far into abstract schemes that bear little resemblance to actual moral practice and that care ethics offers a corrective by insisting on the recognition of turning points—points at which we halt the progression of abstract thought and turn back toward actual life and natural caring.

Consider Marc Hauser's discussion of the evolution of "moral minds."[1] Care ethics agrees with Hauser that human morality has natural roots, but he seeks them along the lines of male experience, whereas I locate them more firmly in female experience. Again, some of us hold with Hauser that morality can be separated entirely from religion (not all care theorists agree on this); we seek natural beginnings. I am perhaps more hostile than Hauser to religion because I believe it has been enormously damaging to women.

But Hauser moves rather too rapidly to claim an ancestral "Rawlsian creature" as an early element in human moral development, and despite a series of powerful arguments attempting to link animal studies to human "intuitions," the general argument just does not hold up. True, some animals at times seem to play by rules, and they seem to know that their self-interest may be advanced by cooperation. But we know that males among our closest nonhuman ancestors were (and still are) often vicious killers. Their cooperation has been based on the protection of their own group from threatening outsiders and an apparent desire to be dominant. Even when animals seem to be acting on rules, something else may be operating. Animals also exhibit maternal care, affection, brotherly (or sisterly) concern, and sorrow.[2] They have feelings. When a mature animal plays gently with a young one, it isn't always clear whether he's playing by rules or showing affection. Thus, I think we should not be hasty in drawing Rawlsian conclusions from animal studies—especially not from laboratory studies.

An even more powerful argument against Hauser's "Rawlsian creature" is the scarcity of Rawlsian humans. Remember that the Rawlsian human arises from a preempirical subject—a highly questionable entity—and it seems odd to claim biological roots for an entity whose empirical roots are explicitly denied. Moreover, we need only look around us to see that many people do not intuitively recognize anything like the difference principle, and some explicitly reject it. Indeed, in the 2008 U.S. election, we heard people reject the principle even when it clearly implied a benefit to themselves. Further, there are those who, asked to play the game behind a veil of ignorance, assert that they would either kill themselves if they turned out to be "one of those people" or fight to become something else. It is likely that most of the people who accept something like the difference principle do so out of care and concern for others, not because they connect fairness to inherent self-interest. A relational self is encumbered from the start.[3] It seems right to seek our moral roots in evolution, but we should recognize a turning point when we move too far from our natural conditions.

There is another set of issues on which Kantians, neo-Kantians, and care theorists might find a bridge across theoretical difference. Kant advised us never to treat another person merely as a means but always as an end. This recommendation is surely accepted by care theorists. But it requires interpretation. How do we follow such a commandment? How do we treat another as an end? For Kant, the categorical imperative leads logically to a set of absolute rules—for example, never deliberately kill, steal, or lie. Carried to its logical extreme, we hear the ancient Father's saying, "I would not tell a willful lie to save the souls of the whole world."[4] In contrast, I (and many other care theorists) would tell a lie to save my pet cat from injury. To treat another as an "end" is to respond (with consideration for the whole web of care) as positively to his needs as possible. For Kant, moral action is rightly determined by consideration of the "moral law within." This moral law has been implanted by God, and for Kant, its presence within us is the only reliable evidence of God's existence. But starting this way ignores our evolutionary past. For care theorists, to treat another as an end does *not* mean to recognize some spark of divinity in him that demands a universalizable response. Indeed, there may be a contradiction in a strict Kantian interpretation. When we refuse to lie no matter who suffers as a result of our refusal, when we refuse to steal when our children are starving, it might be said that we are using these others (or their situations) as a means to our own moral perfection. At the very least, their suffering is regarded as irrelevant to our moral decision making. It seems better, then, to advise that we treat others, not as ends in themselves, but as real, empirical persons with multiple ends. It is these ends to which we must respond with empathy, evaluation, and motivational displacement.

The issue of authority arises dramatically in Kantian ethics. Moral authority, as defined by Kant, rests in the internal moral reason planted in all humans by God. Care ethics sees several problems here. First, there is a worry about any authority anchored in God. Too many people claim special connection to God and do not understand that innate moral *reason* is the authority of which Kant speaks, not some interpretation of God's word. Kant sought to free us—as moral beings—from the authority of both church and state; he tried to lay out a path to moral autonomy. But the connection to God and the ultimate authority given to the "law within" led frequently to an odd detachment from actual empirical life and to an overriding emphasis on duty. That emphasis invites a second difficulty.

Duty has often become unhooked from the authority of the law within and attached to the very authorities Kant wanted to free us from. For care ethics, this is a deeply troubling result.

We do not reject the great moral principles articulated in the last century—for example, the Universal Declaration of Human Rights, the Geneva Conventions, the International Covenant of Civil and Political Rights. The trouble is that national governments interpret and reinterpret such statements to suit their own needs and purposes, and those who should work under these principles find it expedient to obey the interpreting authorities.[5] It is a great heartache to many Americans today that agents of our government have inflicted inhumane, degrading, and humiliating treatment—all forbidden by the principles just mentioned—on many prisoners suspected of being terrorists. Similarly, we are dismayed by the continued use of solitary confinement in our domestic prisons. Prolonged solitary is certainly inhumane, often inducing severe mental illness. Without rejecting the principles as useless, care theorists advise turning away from arguments that concentrate on the wording of principles and abstract interpretations. We must look instead at the situations and particular prisoners at our mercy. Fear, pain, and humiliation should trigger in all moral people an empathic response to relieve suffering, not to inflict it. When authorities or agents try to justify cruel treatment, we should reject the connection between a reinterpretation of principle and the ostensible purpose sought. Because our well-developed empathy will not allow us to deliberately inflict pain, we must find another way. We do not reject the stated purpose—to save the lives of innocents—but we reject methods that cause suffering in our captives and the loss of moral identity in ourselves.

When an empathic response is aroused, we encounter a turning point at which care theorists recognize the limited value of prescriptive principles. Notice that care theorists and Kantians both believe that cruel treatment is wrong. The injunction against such treatment is virtually absolute in both approaches. However, much as care theorists would like to see the great international principles become accurate *descriptive* principles, we do not believe that their establishment as *prescriptive* principles will accomplish this. People have to develop and keep alive their capacity for empathic response. We must be prepared to care.

We share with Utilitarians a concern for people outside the immediate dyad of carer and cared-for. All of our decisions are made with some concern for the "whole web of care." But this web cannot logically extend to the whole world, to dimensions beyond the possibility of caring *relations*, and we do not expect carers to value every life equally or to enhance the good of every child as we do our own. We may wish such good for every child, but we cannot take personal responsibility for providing it. Caring must be completed in the response of the cared-for. We cannot claim for ourselves the power to legislate for all of India from the privacy of our own offices. Still, we recognize the fine intention in considering the greatest good for the greatest number. How can we reasonably approximate this goal? The relevant web of care to which I have regularly referred differs in size from person to person and from time to time, and we risk falling into futile (and sometimes arrogant) patterns of do-goodism when we attempt to stretch the web beyond the point of viability. Instead, in care theory, we try to connect our own webs of care to those already established in other groups, cultures, and nations. We establish links or chains connecting circles of care and work cooperatively to keep open the lines of communication so that we retain some sense of whether *caring-for* is flourishing in each of the webs. We strive for empathic accuracy—understanding is essential—but we try to maintain sympathy even when we cannot condone or support what goes on in alien webs of care. We try to educate sympathetically while remaining educable ourselves.

Utilitarianism moves too far beyond the reality of human living. If, for example, I follow Singer's advice and give 1 percent of my income to reduce starvation and poverty among children in Africa and give nothing to relieve the misery of nonhuman animals in the world, am I guilt of speciesism? I am caught in a dilemma. Again, too much attention goes to the development of principles, computations, and abstract scenarios. Care theory calls for a turning point. Attend to those who address us while carefully observing the effects of our response on others in the web of care. Remain open to the expressed needs of those at a distance and work toward a practical policy for meeting them. We extend our reach, but we stay close to our biological reality.

Care ethics is centered on needs, not rights. As noted above, several powerful thinkers today urge us to address the human suffering that accompanies poverty. They are right to do so, but the strategies they suggest are not always feasible; they are often contrary to human nature. If, say, we decide that hunger is the first need to be met, what should we do? Individuals cannot be expected to forgo all "luxuries" (and what is a luxury?) until there is no hunger in the world. Nor can they be expected to withhold goods from their own children in order to meet the dire needs of others.

Still, moral thinkers should not give up on the problems of needs, however complex and difficult they are. I have suggested that *caring* precedes the identification of needs and that needs are best identified in circles of care. This means that affluent groups who want to help must forge chains of connection with circles of care in suffering groups and take advice from them on priorities and the most promising ways to approach the problems. The first requisite in caring is to listen and be moved.

I have also suggested that the relief of poverty must be a collective effort, not an individual one. Because we are biologically constituted to direct our altruism toward those closest to us, we may feel not only hopelessly confused by the complexity of the problems but also morally uneasy if our efforts deprive those in our circle of natural caring. Again, religion sometimes undermines collective efforts to relieve poverty by making it a sacred duty for individuals to do so. Consider the Catholic Church's reaction to liberation theology: giving to the poor is a Christian duty, but working to remove the causes of their poverty is political work, outside the purview of the church and sometimes condemned by it. It may not be accidental that those nations whose people show the least allegiance to religion give more generously (as nations) to the relief of poverty than those still trapped by institutional authority.

When needs are discussed, the role of caregivers must be addressed. Caring should not be equated with caregiving, and I have emphasized this point again and again. However, caregiving—and the caring professions—should receive a higher place in the economic/occupational world. On this, feminist thinkers from almost all philosophical perspectives agree. In addition to improving the economic conditions of caregivers, however, more attention should be given to the potential of caregiving activities as the incubator of caring. Involving boys as well as girls in such activities may help to develop the empathic capacities that will, in turn, reduce violence at every level. As we agree with liberal philosophers who argue for the equality of women in the public world, care ethicists support the idea that men should become more like women in promoting and sustaining caring relations. Attention to the ethics of care may help to liberate men from the masculine mystique that has enslaved them and cost countless young men their moral identity.

It would be ungrateful and probably hypocritical for any ethical thinker in the Western world to dismiss the liberal notion of individual autonomy. We have come to pride ourselves on making choices without coercion, on being our "own boss." But as I argued in chapter 4, liberal thinkers have built a myth on the concept of autonomy. We are, none of us, completely autonomous. The limited autonomy we enjoy is itself heteronymously derived, dependent on the families and communities into which we are born. It is a thoroughly relational autonomy—acquired in relation and exercised in relation. Moreover, the liberal concept has been further warped toward an emphasis on autonomy as economic and physical self-sufficiency. It is too rarely invoked to challenge the beliefs and practices of the processes of socialization that have formed our supposedly autonomous selves.

Traditional ethical approaches have usually elevated reasoning above feeling. Indeed, Kant insisted that good works or right acts done out of love or inclination have no *moral* worth. Morality is to be associated with locating, accepting, and acting on the correct principle. Some scholars today come close to agreeing with Kant when they argue that morality is a distinctly human capacity anchored in human cognitive ability. Others (with whom I agree) see a continuum between nonhuman and human life, and I have argued that our preferred social condition is characterized by natural caring—a basic, informal moral way of life. The purpose of formal morality, of ethical caring, is to establish or restore natural caring. We are heavily dependent on reasoning whenever we turn to ethical caring.

Possibly no approach to ethics calls more urgently for critical thinking than care ethics. In care ethics, we put little faith in broad, abstract principles. We may give them nominal assent, but we get little direct guidance from them. Nor do we substitute, as Confucians might, a myriad of specific rules to replace universal principles. And we are wary of depending too heavily on our personal virtues as moral agents. We keep our attention on living others to whom we must respond in specific situations. In doing this, we draw on a fund of experience in caring and being cared for, but in every situation we must identify needs, analyze complex interactions, locate similarities and differences between present and past situations, seek empathic accuracy, maintain an open channel between empathy and sympathy, consider the effects of our proposed current response on others in the web of care, and evaluate the resources at our disposal. Contrary to the odd idea that *caring* can be described as a nice, fuzzy feeling, ethical caring requires a high degree of skill in critical thinking, but the required thinking is directed at the situations and practices of real life, not merely at the perfection of theory.

Our schools generally do a poor job in developing students' capacity for the kind of critical thinking I've described—thinking that challenges their own socialization. For example, students are rarely invited to consider the nature and effects of the rituals that promote patriotism. Indeed, although highly educated critics of care theory worry about the parochial effects of caring first within families, they rarely direct similar criticism at national patriotism. But both are anchored in a biological tendency to care for those closest to and most like us. Both should be evaluated with respect and skeptical analysis.

Schools do almost nothing to encourage critical thinking on religion. In the United States, the separation of state and church operates effectively to protect religious institutions from the critical debates that characterize political life. That protection and the consequent lack of critical thinking on religious topics may offer a partial explanation for the continued vigor of religion in the United States while its influence has waned in most of the Western world. Public schools simply cannot engage the kinds of questions I have suggested. Their emphasis (if they give religion any attention at all) is on innocuous bits of information, festivals and customs, and respect for all religions.

But why should women, demeaned and oppressed for centuries by the world's great monotheisms, *respect* religion? Because we care for and about the *people* who embrace religion, we should phrase our criticism with some sensitivity for their feelings. We recognize that many women derive comfort from their religious affiliation and might feel bereft without it. But slowly, patiently, care ethicists should help women to see that religion has been prominent in maintaining their subordination. Religion should not be exempted from critical analysis.

One can separate oneself from institutional religion and retain a belief in God or Gods. One can seek new forms of spirituality. Why do so many people believe in a male deity—

"God the father"? What evidence is there for God's goodness? Would an all-good God liberate his favored people by killing all the first-borns of their captors? Would an all-good God permit anyone to suffer in hell for eternity? It will be a day of genuine liberation when women insist upon an apology from the religious authority that has so oppressed them. If the apology is not forthcoming, they should reject institutional religion and feel comfortable in devoting their caring to life on and for this earth. We can admit with genuine sympathy that the loss of institutional religion creates a nostalgia for old hymns, rituals, prayers, and communion. But once free, those who wish to do so may enter a new phase of spiritual seeking. Many who have abandoned religion nevertheless experience awe at the grandeur of the universe, the diversity of life, and the wonder of human love.

Care ethics, like virtue ethics, relies heavily on factors internal to moral agents—to those who care. It supports the development of virtues, but like Socrates it is skeptical about identifying virtues in the abstract and attempting to inculcate them directly. Virtues develop in ways of life, and it should not be surprising that so many highly admired virtues have been associated with military life. The long association of virtue with masculinity has warped our whole system of valuation, and virtues associated with female experience have long been undervalued.

Care ethics argues that virtues are developed in situational practices. For example, caregiving often acts as an incubator of caring. If we want children to develop virtues associated with caring, we must first care for them and help them to understand what it means to be cared for. Then we must give both girls and boys opportunities to engage in caregiving. Central to this project of moral education are dialogue, practice, and confirmation. We listen and talk, explore possibilities, and practice together. We point each child toward the child's better self. Preaching is relatively ineffective, and that is why many forms of character education miscarry. Stories can help by setting the stage and providing exemplars, but the virtues themselves emerge in practice as *virtuous acts*. Situations, interactions, and expressed needs call forth virtuous responses. One does not gird oneself with virtues inculcated by elders and then go forth to find situations in which to practice them. The stored memories of caring and being cared for act as reservoirs for the production of virtuous acts, and since this reservoir is built in relation, it is unlikely that the resulting virtues will be considered individual possessions. Carers will abandon a specific virtue just as they abandon an abstract principle if the expressed need of another is convincingly urgent and if its satisfaction will not harm others in the web of care.

Care ethics also has much in common with the so-called sentimentalists—those philosophers who argued that emotion, not reason, is the motivational force in moral decision making. We do not claim, however, that goodness—say, benevolence—is somehow built into human beings by God. Starting with a natural phenomenon, maternal instinct, we trace the development of empathy, mutuality, natural caring, and ethical caring. Whereas Hume had no point of origin on which to rest his claim of an innate tendency toward human benevolence, today's care theorists can point to evolutionary evidence of other-centered emotions such as maternal instinct and group bonding.

In many of the topics considered in this book—the use of principles, exercise of virtue, possibility of autonomy, and the moral duty to meet needs—there is some convergence between traditional ethics and the ethics of care. In each area, the female thinker puts a restraining hand on the philosophical arm of her male companion: Wait! Don't go too far. Come back to earth and consider real people in real situations. Above all, recognize the relational nature of human reality.

In the matter of war, however, the possibility of convergence is very small. The force of patriarchy and masculinity in history, religion, patriotic rituals, stories, laws, and songs is enormous, and women—loyal in their subordination—too easily and too often join the ranks of the militant or passive. From the perspective of care ethics, war is immoral; it is nonsense to talk about the moral conduct of war. When it becomes the patriotic duty of one group of men to kill members of another group, destroy their homes and livelihoods, inflict terror ("awe"?) on their countrymen, morality vanishes. It belongs to a different world.

Care ethics, bound so closely to this earth, has clear affinity with peace groups but cannot logically advocate absolute pacifism. We know that we will fight to preserve our own lives and those of our children. Our moral obligation is to prevent as nearly as possible the situations in which fighting becomes necessary. This is an argument that Sara Ruddick has made also.[6] When individuals or groups resort to violence, force—legal, cooperatively endorsed force—should be used to stop the violence. But this force should not be called "war." The United States has not been engaged in war in either Iraq or Afghanistan; there is no conflict between nations. Refusing the label "war" to such engagements would reduce the power of those who instigate the violence. They would be branded by world opinion as criminals. Without the label "war," there would be no patriotic duty to fight for one's country, defend one's flag, or fight anew so that those already dead will not have died in vain. Instead of invoking national patriotism, we might work toward a new vision—establishment of a cooperatively sanctioned force to preserve life and peace. Even that force would have to be monitored closely. We might find a way to appeal to civic-minded young people to serve in this force—not for glory or to experience the "delight of destruction"—but as a contribution to world peace.

There is clearly much more work for care ethicists to tackle. Virginia Held has argued that there should be much more cooperation among the various domains—political, economic, cultural—to which care ethics has made some contribution.[7] We need to say more about a care-driven theory of justice, restorative justice, and the possibilities of care in the economic domain. Ruth Groenhout has also suggested that care ethicists need to say something about the origins of evil and appropriate responses to it.[8] Provided that we are not distracted by notions of cosmic evil and how it invaded an otherwise perfect world, I think she is right; that is, we need to say more about possible ways to respond to deceit, greed, and aggression.

In this book, I have attempted to broaden and deepen the discussion of care ethics and to address questions that have arisen in the past twenty years. Instead of backing away from the identification of care ethics as a woman's approach to moral life, I have embraced that identity, located its origins in maternal instinct, and anchored it in female experience. This is not to say that men cannot share in this ethic. They can and should do so. Male-proud traditions have defined masculinity in such a way that boys have little opportunity to develop their tender, nurturing tendencies. Indeed, such tendencies in boys have been mocked and scorned.

My position on biological differences will bother, even anger, some readers. Those who believe that all differences along gender lines are the product of culture and socialization argue that the scarcity of women at the highest levels of mathematics, science, engineering, and some of the arts can be blamed on lack of opportunity and prejudice. They may be right. They are certainly at least partly right. However, it would be astonishing if thousands of years of female responsibility for the care of others did not leave an evolutionary mark. Recognizing biological/evolutionary factors leads logically and practically to a serious program of reevaluating values, virtues, and practices. Women should, of course, have opportunities to participate fully in the male-defined world, but that definition should change. Not everything should be judged by an established male standard. A reevaluation of the entire field of human relations should be undertaken to infuse caregiving, diplomacy, global interaction, and family practices with an appreciation of the attitudes, skills, and understanding described in care ethics. In this area of human activity, the best female models and practices should set the standards. In undertaking such a project, we should remember that our own standards should come under continual analysis; care theory itself may be described differently by different thinkers, and there is still much work to be done.

NOTES

INTRODUCTION

1. See, for example, Donald M. Broom, *The Evolution of Morality and Religion* (Cambridge: Cambridge University Press, 2003); Frans de Waal, *Primates and Philosophers: How Morality Evolved* (Princeton, NJ: Princeton University Press, 2006); Lee Alan Dugatkin, *The Altruism Equation* (Princeton, NJ: Princeton University Press, 2006); Marc D. Hauser, *Moral Minds* (New York: Harper Collins, 2006); and Richard Joyce, *The Evolution of Morality* (Cambridge, MA: MIT Press, 2006).

2. Virginia Held, "Moral Subjects: The Natural and the Normative," *Proceedings and Addresses of the American Philosophical Association* 76, no. 2 (2002): 7–24.

3. I made a start on this project earlier. See Nel Noddings, *Caring: A Feminine Approach to Ethics and Moral Education* (Berkeley and Los Angeles: University of California Press, 1984).

4. "Girls Make History," *New York Times*, December 4, 2007, B-1.

5. Michael Slote has developed a form of care ethics based entirely on the notion of empathy. See Slote, *The Ethics of Care and Empathy* (New York: Routledge, 2007).

CHAPTER 1. THE EVOLUTION OF MORALITY

1. See Martin Hoffman, *Empathy and Moral Development: Implications for Caring and Justice* (New York: Cambridge University Press, 2000); and Michael Slote, *The Ethics of Care and Empathy* (New York: Routledge, 2007).

2. Hoffman, *Empathy and Moral Development*, 30.

3. Donald M. Broom, *The Evolution of Morality and Religion* (Cambridge: Cambridge University Press, 2003), 220.

4. See Karsten R. Stueber, *Rediscovering Empathy* (Cambridge, MA: MIT Press, 2006); see also Susan Verducci, "A Conceptual History of Empathy and a Question It Raises for Moral Education," *Educational Theory* 50, no. 1 (2001): 63–80.

5. Nel Noddings, *Starting at Home: Caring and Social Policy* (Berkeley and Los Angeles: University of California Press, 2002), 13–14.

6. See, for example, Lee Alan Dugatkin, *The Altruism Equation* (Princeton, NJ: Princeton University Press, 2006).

7. Frans de Waal, *Primates and Philosophers: How Morality Evolved* (Princeton, NJ: Princeton University Press, 2006).

8. Marc D. Hauser, *Moral Minds* (New York: Harper Collins, 2006), 7.

9. See Richard Joyce, *The Evolution of Morality* (Cambridge, MA: MIT Press, 2006), for a discussion of these beginnings. However, after mentioning the importance of mother-love, he quickly moves on to other matters. His move from instinctive mother-love to parental love is especially problematic.

10. For a discussion of relevant neurological and hormonal differences between the sexes, see Sarah Blaffer Hrdy, *Mothers and Others* (Cambridge, MA: Belknap Press of Harvard University Press, 2009).

11. See Joyce, *Evolution of Morality*, on this. For an argument against the complete naturalization of care ethics, see Virginia Held, "Moral Subjects: The Natural and the Normative," *Proceedings and Addresses of the American Philosophical Association* 76, no. 2 (2002): 7–24.

12. See Alasdair MacIntyre, *After Virtue* (Notre Dame, IN: University of Notre Dame Press, 1981).

13. Carol J. Gilligan, *In a Different Voice* (Cambridge, MA: Harvard University Press, 1982).

14. For a helpful summary of the work on care ethics, see Virginia Held, *The Ethics of Care: Personal, Political, and Global* (Oxford: Oxford University Press, 2006).

15. See Immanuel Kant, "Of the Distinction Between the Beautiful and the Sublime in the Interrelations of the Two Sexes," in *Philosophy of Woman*, ed. Mary Briody Mahowald (Indianapolis: Hackett, 1983), 194.

16. See Lawrence Kohlberg, *The Philosophy of Moral Development*, vol. 1 (San Francisco: Harper & Row, 1981).

17. For reactions in the professions, see Nel Noddings, "Feminist Critiques in the Professions," in *Review of Research in Education*, vol. 16, ed. Courtney B. Cazden (Washington, DC: American Educational Research Association, 1990), 393–424.

18. See Annette Baier, *Moral Prejudices: Essays on Ethics* (Cambridge, MA: Harvard University Press, 1994).

19. Rosalind Barnett and Caryl Rivers do this and even suggest, wrongly, that the idea itself can be traced to Gilligan. See Barnett and Rivers, *Same Difference: How Gender Myths Are Hurting Our Relationships, Our Children, and Our Jobs* (New York: Basic Books, 2004).

20. See the essays in Deborah L. Rhode, ed., *Theoretical Perspectives on Sexual Difference* (New Haven, CT: Yale University Press, 1990).

21. See Barnett and Rivers, *Same Difference.*

22. Slote, *Ethics of Care and Empathy*, 1.

23. On the mistakes of utopian thinking, see John Gray, *Black Mass: Apocalyptic Religion and the Death of Utopia* (Canada: Doubleday, 2007).

24. I argued this in Noddings, *Starting at Home.*

25. Susan Moller Okin has made a convincing argument that caring, as care theorists have described it, is implicitly incorporated in John Rawls's theory of justice. See Okin, "Reason and Feeling in Thinking about Justice," *Ethics* 99, no. 2 (1989): 229–49.

26. I discussed the need for turning points in Noddings, *Caring: A Feminine Approach to Ethics and Moral Education* (Berkeley and Los Angeles: University of California Press, 1984).

27. Noddings, *Caring*, 109–12.

28. See Noddings, *Starting at Home*.

29. See, for example, Frances H. Early, *A World without War* (Syracuse, NY: Syracuse University Press, 1997); Jean Bethke Elshtain, *Women and War* (New York: Basic Books, 1987); and Sara Ruddick, *Maternal Thinking: Toward a Politics of Peace* (Boston: Beacon Press, 1989).

CHAPTER 2. THE CARING RELATION

1. See the summary in Virginia Held, *The Ethics of Care: Personal, Political, and Global* (Oxford: Oxford University Press, 2006).

2. On maternal thinking, see Sara Ruddick, *Maternal Thinking: Toward a Politics of Peace* (Boston: Beacon Press, 1989).

3. See Philippa Foot, *Natural Goodness* (Oxford: Oxford University Press, 2001).

4. See Daniel Lord Smail, *On Deep History and the Brain* (Berkeley and Los Angeles: University of California Press, 2007).

5. Ibid.

6. Jean Piaget, *Biology and Knowledge* (Chicago: University of Chicago Press, 1971), 368.

7. Carol J. Gilligan, *In a Different Voice* (Cambridge, MA: Harvard University Press, 1982).

8. Marc D. Hauser, *Moral Minds* (New York: Harper Collins, 2006), xviii.

9. See Catherine A. MacKinnon, *Feminism Unmodified* (Cambridge, MA: Harvard University Press, 1987). But on neurogenic and hormonal differences, see Sarah Blaffer Hrdy, *Mothers and Others* (Cambridge, MA: Belknap Press of Harvard University Press, 2009).

10. See, for example, Michael Slote, *The Ethics of Care and Empathy* (New York: Routledge, 2007).

11. See Martin Buber, *I and Thou*, trans. Walter Kaufmann (New York: Charles Scribner's Sons, 1970). I also discuss unequal relations at some length in Nel Noddings, *Caring: A Feminine Approach to Ethics and Moral Education* (Berkeley and Los Angeles: University of California Press, 1984).

12. For discussion of the role of mothering in the "care trap," see Nancy Chodorow, *The Reproduction of Mothering* (Berkeley and Los Angeles: University of California Press, 1978).

13. Simone Weil, *Simone Weil Reader*, ed. George A. Panichas (Mt. Kisco, NY: Moyer Bell Limited, 1977), 51.

14. See Noddings, *Caring*.

15. Buber, *I and Thou*, 59.

16. Weil, *Simone Weil Reader*, 44–52.

17. Ibid., 313–39.

18. Iris Murdoch, *The Sovereignty of Good* (London: Routledge & Kegan Paul, 1970), 28.

19. William W. Ickes writes about empathic accuracy. See his *Empathic Accuracy* (New York: Guilford, 1997).

20. Murdoch, *Sovereignty of Good*, 17.

21. Ibid., 23.

22. Michael Stocker, *Valuing Emotions* (Cambridge: Cambridge University Press, 1996), 216–17.

23. See the chapters on Murdoch (chaps. 1 and 2) in Lawrence A. Blum, *Moral Perception and Particularity* (Cambridge: Cambridge University Press, 1994).

24. This possibility is described in some depth in Ruddick, *Maternal Thinking*. However, Ruddick considers men as well as women capable of maternal thinking. See also Hrdy, *Mothers and Others*, on this.

25. See Jon Entine, *Abraham's Children: Race, Identity, and the DNA of the Chosen People* (New York: Grand Central, 2007); also Sean B. Carroll, *The Making of the Fittest* (New York: W.W. Norton, 2006).

26. I have described such cases more fully in Noddings, *Starting at Home: Caring and Social Policy* (Berkeley and Los Angeles: University of California Press, 2002), using as illustrative the case of Theobald and

Ernest in Samuel Butler's *The Way of All Flesh* (Garden City, NY: Doubleday, 1944).

27. Martin Hoffman, *Empathy and Moral Development* (New York: Cambridge University Press, 2000), 143.

28. Ruddick, *Maternal Thinking*, 104.

29. See Ruddick, *Maternal Thinking*, for a wide-ranging discussion of the issues that arise in dialogue.

30. James Terry White, *Character Lessons in American Biography* (New York: Character Development League, 1909).

31. See Hoffman, *Empathy*, on induction and transgressive guilt.

32. Hoffman, *Empathy*, 119.

33. For more on confirmation in moral education, see Noddings, *Educating Moral People* (New York: Teachers College Press, 2002).

CHAPTER 3. ETHICAL CARING AND OBLIGATION

1. Michael Slote, *The Ethics of Care and Empathy* (New York: Routledge, 2007), 43.

2. See the study of Holocaust rescuers in Samuel P. Oliner and Pearl M. Oliner, *The Altruistic Personality: Rescuers of Jews in Nazi Europe* (New York: The Free Press, 1988).

3. Virginia Held, *The Ethics of Care: Personal, Political, and Global* (Oxford: Oxford University Press, 2006), 12.

4. Margaret Urban Walker, "Moral Understandings: Alternative 'Epistemology' for a Feminist Ethics," *Hypatia* 4 (Summer 1989): 20.

5. For a helpful summary, see Held, *Ethics of Care*.

6. Eva Feder Kittay also makes this distinction in her influential book *Love's Labor: Essays on Women, Equality, and Dependency* (New York: Routledge, 1999).

7. See Sara Ruddick, *Maternal Thinking: Toward a Politics of Peace* (Boston: Beacon Press, 1989).

8. See, for example, Lucy Candib, *Medicine and the Family: A Feminist Perspective* (New York: Basic Books, 1995); Kari Waerness, "The Rationality of Caring," *Economic and Industrial Democracy* 5, no. 2 (1984): 185–212; and Jean Watson, *Nursing: Human Science and Human Care* (Norwalk, CT: Appleton-Century-Crofts, 1985).

9. See Susan Reverby, *Ordered to Care* (Cambridge: Cambridge University Press, 1987).

10. See Jane Roland Martin, *Reclaiming a Conversation* (New Haven, CT: Yale University Press, 1985).

11. See Linda Babcock and Sara Laschever, *Women Don't Ask: Negotiation and the Gender Divide* (Princeton, NJ: Princeton University Press, 2003).

12. See Deborah L. Rhode, "Definitions of Difference," in *Theoretical Perspectives on Sexual Difference*, ed. Deborah L. Rhode (New Haven, CT: Yale University Press, 1990), 197–212.

13. See Mary K. Zimmerman, Jacqueline S. Litt, and Christine E. Bose, eds., *Global Dimensions of Gender and Carework* (Stanford, CA: Stanford University Press, 2006); also Suzanne M. Bianchi, Lynne M. Casper, and Rosalind Berkowitz King, eds., *Work, Family, Health, and Well-Being* (Mahwah, NJ: Lawrence Erlbaum, 2005).

14. Anne Morrow Lindbergh, *Gift from the Sea* (New York: Random House, 1955), 124–25.

15. Peter Singer, *One World: The Ethics of Globalization* (New Haven, CT: Yale University Press, 2002), 194.

16. On this, see Ruddick, *Maternal Thinking*.

17. Singer, *One World*, 189.

18. See, for example, Lee Alan Dugatkin, *The Altruism Equation* (Princeton, NJ: Princeton University Press, 2006).

19. Peter Unger, *Living High and Letting Die* (New York: Oxford University Press, 1996).

20. Slote, *Ethics of Care and Empathy*, 21–37; see also Slote, "Caring in the Balance," in *Norms and Values: Essays on the Work of Virginia Held*, ed. Joram G. Haber and Mark S. Halfon (Lanham, MD: Rowman & Littlefield, 1998): 27–36.

21. See Slote, "Famine, Affluence, and Virtue," in *Working Virtue*, ed. Rebecca L. Walker and Philip J. Ivanhoe (Oxford: Oxford University Press, 2007): 279–96.

22. Singer, *One World*, 194. Thomas Pogge also estimates that the worst poverty could be eliminated if wealthy nations gave 1.2 percent of their income to developing nations. But notice, again, that such a move would have to be embedded in a much larger program that would address fundamental institutional structures and changes necessary to achieve sustainability. See Pogge, *World Poverty and Human Rights* (Cambridge: Polity/Blackwell, 2002).

23. Slote, *Ethics of Care and Empathy*, 33.

24. See Alasdair MacIntyre, *Whose Justice? Which Rationality?* (Notre Dame, IN: Notre Dame University Press, 1988).

25. See Michael S. Katz, Nel Noddings, and Kenneth A. Strike, eds., *Justice and Caring: The Search for Common Ground in Education* (New York: Teachers College Press, 1999).

26. Held, *Ethics of Care*, 15.

27. John Rawls makes self-interest central in his use of the veil of ignorance and the original position. See Rawls, *A Theory of Justice* (Cambridge, MA: Harvard University Press, 1971). But see the interpretation involving care by Susan Moller Okin, "Reason and Feeling in Thinking about Justice," *Ethics* 99, no. 2 (1989): 229–49.

28. Joseph Stiglitz, *Globalization and Its Discontents* (New York: W.W. Norton, 2002), xvi.

29. Harold Saunders, a former Assistant Secretary for Near Eastern and South Asian Affairs, has written persuasively along these lines. See Harold H. Saunders, *The Other Walls* (Princeton, NJ: Princeton University Press, 1991).

30. See Edward O. Wilson, *The Creation: An Appeal to Save Life on Earth* (New York: W.W. Norton, 2006).

31. Ibid., 168.

32. See Fiona Robinson, *Globalizing Care* (Boulder, CO: Westview Press, 1999).

33. Pogge, *World Poverty and Human Rights*, 70.

CHAPTER 4. THE LIMITS OF AUTONOMY

1. Martha Fineman, *The Autonomy Myth: A Theory of Dependency* (New York: New Press, 2004).

2. Ibid., 20.

3. Ibid., 285.

4. See the accounts in Suzanne M. Bianchi, Lynee M. Cooper, and Rosalind Berkowitz King, eds., *Work, Family, Health, and Well-Being* (Mahwah, NJ: Lawrence Erlbaum, 2005).

5. George Orwell, *Nineteen Eighty-Four* (New York: Harcourt, Brace, & World, 1949), 176.

6. Kevin Tierney, *Darrow: A Biography* (New York: Thomas Y. Crowell, 1979), 340. Tierney cites the source as Nathan Freudenthal Leopold's 1958 book, *Life Plus 99 Years* (New York: Popular Library).

7. Tierney, *Darrow*, 341.

8. See Jean-Paul Sartre, *Being and Nothingness*, trans. Hazel E. Barnes (New York: Washington Square Press, 1956); and Sartre, *Nausea*, trans. Lloyd Alexander (Norfolk, CT: New Directions, 1959).

9. Viktor E. Frankl, *The Doctor and the Soul* (New York: Vintage Books, 1973).

10. See my discussion of suffering and choice in Noddings, *Happiness and Education* (Cambridge: Cambridge University Press, 2003).

11. Orwell, *Nineteen Eighty-Four*, 239.

12. John Rawls, *A Theory of Justice* (Cambridge, MA: Harvard University Press, 1971). See also the critique by Michael Sandel, *Liberalism and the Limits of Justice* (Cambridge: Cambridge University Press, 1982).

13. Quoted in Virginia Held, *The Ethics of Care: Personal, Political, and Global* (Oxford: Oxford University Press, 2006), 13. Original quotation from Martha Nussbaum, *Sex and Social Justice* (New York: Oxford University Press, 1999), 62.

14. See the essays in Catriona Mackenzie and Natalie Stoljar, eds., *Relational Autonomy* (Oxford: Oxford University Press, 2000).

15. On the likely conflicts of interest, see Grace Clement, *Care, Autonomy, and Justice: Feminism and the Ethic of Care* (Boulder, CO: Westview Press, 1996).

16. Held, *Ethics of Care*, 55; Martin Buber, *I and Thou*, trans. Walter Kaufmann (New York: Charles Scribner's Sons, 1970).

17. See James Gilligan, *Violence* (New York: G. P. Putnam's Sons, 1996).

18. Ibid., 267.

19. Samuel Butler, *The Way of All Flesh* (Garden City, NY: Doubleday, 1944).

20. Diana T. Meyers has written extensively and persuasively about the need to examine our socialization critically. See Meyers, *Self, Society, and Personal Choice* (New York: Columbia University Press, 1989). See also the discussion in Mackenzie and Stoljar, *Relational Autonomy*, 124–50.

21. For readers interested in debates about the meaning of critical thinking and how to teach it, see Noddings, *Critical Lessons: What Our Schools Should Teach* (Cambridge: Cambridge University Press, 2006).

22. The value of carework, usually unpaid at home, is discussed appreciatively in Riane Eisler, *The Real Wealth of Nations* (San Francisco: Berrett-Koehler, 2007).

23. Sonya Michel, *Children's Interests/Mothers' Rights* (New Haven, CT: Yale University Press, 1999), 261.

24. See Eisler, *Real Wealth of Nations*.

25. Quoted by Sandra Tsing Loh, "I Choose My Choice!" *Atlantic*, July/August, 2008, 127, in a review of Linda Hirshman, *Get to Work . . . and Get a Life, Before It's Too Late* (New York: Penguin, 2007).

26. Edward Bellamy, *Looking Backward* (New York: New American Library, 1960/1897), 90.

27. This information was forwarded to me as part of an ongoing conversation with Alfie Kohn, who is challenging the notion that self-discipline is an indisputable virtue.

CHAPTER 5. RELATION, VIRTUE, AND RELIGION

1. Martin Buber, *I and Thou*, trans. Walter Kaufmann (New York: Charles Scribner's Sons, 1970), 69.

2. Virginia Held, *The Ethics of Care: Personal, Political, and Global* (Oxford: Oxford University Press, 2006), 46.

3. See the account in Held, *Ethics of Care*.

4. See, for example, Alasdair MacIntyre, *After Virtue* (Notre Dame, IN: University of Notre Dame Press, 1981); MacIntyre, *Whose Justice? Which Rationality?* (Notre Dame, IN: Notre Dame University Press, 1988); Charles Taylor, *Sources of the Self* (Cambridge, MA: Harvard University Press, 1989).

5. Buber, *I and Thou*, 67.

6. From a personal communication with Michael Slote on his new book, *Moral Sentimentalism* (Oxford: Oxford University Press, 2010).

7. See Noddings, "Caring as Relation and Virtue in Teaching," in *Working Virtue*, ed. Rebecca L Walker and Philip J. Ivanhoe (Oxford: Oxford University Press, 2007), 41–60.

8. See the criticisms in the reviews in a symposium entitled *Caring* in *Hypatia* 5, no. 1 (1990): 101–26.

9. Nel Noddings, *Caring: A Feminine Approach to Ethics and Moral Education* (Berkeley and Los Angeles: University of California Press, 1984), 109–11.

10. Raja Halwani, "Care Ethics and Virtue Ethics," *Hypatia* 18, no. 3 (2003): 170.

11. Ibid., 187.

12. See Chenyang Li, "The Confucian concept of Jen and the Feminist Ethics of Care: A Comparative Study," *Hypatia* 9, no. 1 (1994): 70–89.

13. See Daniel Star, "Do Confucians Really Care? A Defense of the Distinctiveness of Care," *Hypatia* 17, no. 1 (2002): 77–106.

14. Ibid.

15. See the discussion in Star, "Do Confucians Really Care?" Also Margery Wolf, "Beyond the Patrilineal Self: Constructing Gender in China," in *Self as Person in Asian Theory and Practice*, ed. Roger T. Ames, Wimal Dissanayake, and Thomas P. Kasulis (Albany: State University of New York Press, 1994).

16. Ruth E. Groenhout, *Connected Lives: Human Nature and an Ethics of Care* (Lanham, MD: Rowman & Littlefield, 2004).

17. For a powerful argument against the notion that God is demonstrably good and loving, see Bart D. Ehrman, *God's Problem* (New York: HarperCollins, 2008).

18. The comments appear in Yossi Klein Halevi, *At the Entrance to the Garden of Eden* (New York: Perennial, 2001), 254–55.

19. Simon Wiesenthal, *The Sunflower* (New York: Schocken Books, 1976).

20. See Richard Dawkins, *The God Delusion* (Boston: Houghton Mifflin, 2006).

21. See Paul Bloom, *Descartes' Baby* (New York: Basic Books, 2004).

22. David L. Linden, *The Accidental Mind* (Cambridge, MA: Belknap Press of Harvard University Press, 2007), 225.

23. See my discussion in Noddings, "The New Outspoken Atheism and Education," *Harvard Educational Review* (Summer 2008): 369–90.

24. Martin Gardner, *The Whys of a Philosophical Scrivener* (New York: Quill, 1983), 352.

25. Daniel Dennett, *Breaking the Spell* (New York: Viking, 2006), 210.

26. Paul Ricoeur, *The Symbolism of Evil* (Boston: Beacon Press, 1969), 239.

27. Mary Daly, *Beyond God the Father* (Boston: Beacon Press, 1974), 195.

28. Ibid., 3.

29. John Anthony Phillips, *Eve: The History of an Idea* (San Francisco: Harper & Row, 1984), 174.

30. Dawkins, *God Delusion*, 31.

31. Virginia Woolf, "Professions for Women," in *Collected Essays*, vol. 2 (London: Hogarth Press, 1966), 285.

32. Many references are available, among them Christine Downing, *The Goddess* (New York: Crossroad, 1984); Riane Eisler, *The Chalice and the Blade* (New York: Harper Collins, 1987); and Merlin Stone, *When God was a Woman* (New York: Dial Press, 1976).

33. See Christian Smith, *Soul Searching: The Religious and Spiritual Lives of American Teenagers* (Oxford: Oxford University Press, 2005).

34. Charles Kimball, *When Religion Becomes Evil* (New York: Harper Collins, 2002), 39.

CHAPTER 6. EMOTIONS AND REASON

1. Michael Slote, *The Ethics of Care and Empathy* (New York: Routledge, 2007), 3.

2. Annette Baier, *Moral Prejudices: Essays on Ethics* (Cambridge, MA: Harvard University Press, 1994).

3. David Hume, *An Enquiry Concerning the Principles of Morals* (Indianapolis: Hackett, 1983/1751), 15.

4. Alasdair MacIntyre, *After Virtue* (Notre Dame, IN: University of Notre Dame Press, 1981), 49.

5. An impressive volume of literature is available on this topic. See, for example, Marc Bekoff, *Minding Animals* (Oxford: Oxford University Press, 2002); David DeGrazia, *Taking Animals Seriously* (Cambridge: Cambridge University Press, 1996); Jeffrey Moussaieff Masson and Susan McCarthy, *When Elephants Weep: The Emotional Lives of Animals* (New York: Delacorte Press, 1995); Frans de Waal, *Primates and Philosophers: How Morality Evolved* (Princeton, NJ: Princeton University Press, 2006).

6. Hume, *Enquiry*, 18.

7. See Jane Goodall's foreword to Bekoff, *Minding Animals*.

8. Jeremy Bentham, *Introduction to the Principles of Morals and Legislation* (Oxford: Clarendon Press, 1996/1789), chap. 17.

9. Jonathan H. Turner, *On the Origins of Human Emotions* (Stanford, CA: Stanford University Press, 2000), 120.

10. De Waal, *Primates and Philosophers*, 4.

11. Ibid., 5.

12. Samuel P. Oliner and Pearl M. Oliner, *The Altruistic Personality: Rescuers of Jews in Nazi Europe* (New York: The Free Press, 1988), 257.

13. Ibid.

14. See Noddings, "On Community," *Educational Theory* 46, no. 3 (1996): 245–67.

15. Martha C. Nussbaum, *Sex and Social Justice* (New York: Oxford University Press, 1999), 74.

16. Quoted by Nussbaum, *Sex and Social Justice*, 75. From Noddings, *Caring: A Feminine Approach to Ethics and Moral Education* (Berkeley and Los Angeles: University of California Press, 1984), 137.

17. Nussbaum, *Sex and Social Justice*, 76.

18. Ibid., 75.

19. Noddings, *Caring*, 5.

20. For studies of somatic matching of emotional states, see Antonio Damasio, *Descartes' Error: Emotion, Reason, and the Human Brain* (New York: Putnam, 1994).

21. De Waal, *Primates and Philosophers*, 6.

22. For a detailed and sophisticated discussion of the history of virtue and virtues, see MacIntyre, *After Virtue;* see also, especially on courage, Paul Tillich, *The Courage to Be* (New Haven, CT: Yale University Press, 1952).

23. See James Terry White, *Character Lessons in American Biography* (New York: Character Development League, 1909).

24. John Rawls, *A Theory of Justice* (Cambridge, MA: Harvard University Pres, 1971), 192.

25. Sara Ruddick, *Maternal Thinking: Toward a Politics of Peace* (Boston: Beacon Press, 1989), 79.

26. MacIntyre, *After Virtue*, 191.

27. Hume, *Enquiry*, 61–72.

28. Ibid., 73.

29. Ibid., 74.

30. Daniel C. Dennett, *Breaking the Spell* (New York: Viking, 2006), 306.

31. Ibid.

CHAPTER 7. NEEDS, WANTS, AND INTERESTS

1. See the discussions in David Braybrooke, *Meeting Needs* (Princeton, NJ: Princeton University Press, 1987); and Joan Tronto, *Moral Boundaries: A Political Argument for an Ethic of Care* (New York: Routledge, 1993).

2. Tronto, *Moral Boundaries*, 138.

3. Ibid., 127.

4. See Braybrooke, *Meeting Needs*, chap. 8.

5. Abraham H. Maslow, *Motivation and Personality* (New York: Harper & Row, 1970). For higher forms of need satisfaction, see Maslow, *The Farther Reaches of Human Nature* (New York: Viking Press, 1971).

6. See, for example, the accounts in Primo Levi, *The Drowned and the Saved*, trans. Raymond Rosenthal (New York: Vintage, 1988).

7. See Braybrooke, *Meeting Needs*, chap. 8.

8. See the account in Michael True, *An Energy Field More Intense Than War* (Syracuse, NY: Syracuse University Press, 1995).

9. See, for example, the responses by Sarah Lucia Hoagland ("Some Concerns about Nel Noddings' *Caring*"), Claudia Card ("Caring and Evil"), and Barbara Houston ("Caring and Exploitation") in the symposium on *Caring* in *Hypatia* 5, no. 1 (1990): 101–19; and my answer to this charge, Noddings, "On the Alleged Parochialism of Caring," *Newsletter on Feminism* (American Philosophical Association, 1991), 96–99.

10. Sara Ruddick, *Maternal Thinking: Toward a Politics of Peace* (Boston: Beacon Press, 1989).

11. John Dewey, *Experience and Education* (New York: Macmillan, 1963), 67.

12. Daniel J. Levitin, *The World in Six Songs* (New York: Dutton, 2008), 62.

13. Mattathias Schwartz, "Mal*Web*olence," *New York Times Magazine*, August 3, 2008, 24–29.

14. See Noddings, *Critical Lessons: What Our Schools Should Teach* (Cambridge: Cambridge University Press, 2006).

15. Isaiah Berlin, *Four Essays on Liberty* (Oxford: Oxford University Press, 1969).

CHAPTER 8. WAR AND VIOLENCE

1. See John Howard Yoder, *The Politics of Jesus* (Grand Rapids, MI: Eerdmans, 1994).

2. Daniel J. Levitin, *The World in Six Songs* (New York: Penguin, 2008), 65.

3. Michael Walzer, *Just and Unjust Wars* (New York: Basic Books, 1977), 36.

4. See Jonathan Glover, *Humanity: A Moral History of the 20th Century* (New Haven, CT: Yale University Press, 2000); J. G. Gray, *The Warriors: Reflections on Men in Battle* (Lincoln, NE: Bison Press, 1998); Peter Schrijvers, *The GI War against Japan* (New York: New York University Press, 2002).

5. Walzer, *Just and Unjust Wars*, 44.

6. Ibid., 335.

7. See Jean Bethke Elshtain, *Jane Addams and the Dream of American Democracy* (New York: Basic Books, 2002).

8. Anthony Swofford, *Jarhead* (New York: Scribner, 2003), 6–7.

9. Dexter Filkins, *The Forever War* (New York: Knopf, 2008), 296–306.

10. Chris Hedges, "War Is a Force That Gives Us Meaning," *Amnesty Now*, (Winter 2002): 10–13.

11. Gray, *The Warriors*, 25–58.

12. Ibid., 52.

13. For a description of the male tendency to aggression and violence, see Richard Wrangham and Dale Peterson, *Demonic Males: Apes and the Origins of Human Violence* (New York: Houghton Mifflin, 1996).

14. William James, *The Varieties of Religious Experience* (New York: Modern Library, 1929/1902), 359. See also James, "The Moral Equivalent of War," in *The Writings of William James*, ed. John J. McDermott (New York: Random House, 1967), 660–671.

15. Wrangham and Peterson, *Demonic Males*, 113.

16. See Emilie Buchwald, Pamela R. Fletcher, and Martha Roth, eds., *Transforming a Rape Culture* (Minneapolis: Milkweed Editions, 1993).

17. Clark Wissler, *Indians of the United States* (New York: Anchor Books, 1989), 242.

18. Levitin, *World in Six Songs*, 45.

19. Ibid., 289.

20. See Jean Bethke Elshtain, *Women and War* (New York: Basic Books, 1987).

21. Ibid., 234.

22. There are many useful accounts, for example, Birgit Brock-Utne, *Educating for Peace: A Feminist Perspective* (New York: Pergamon, 1985); Cambridge Women's Peace Collective, *My Country Is the Whole World: An Anthology of Women's Work on Peace and War* (Boston: Pandora Press, 1984); David Cortright, *Peace: A History of Movements and Ideas* (Cambridge: Cambridge University Press, 2008); Frances H. Early, *A World without War* (Syracuse, NY: Syracuse University Press, 1997); Jean Bethke Elshtain and Sheila Tobias, eds., *Women, Militarism, and War* (Savage, MD: Rowman & Littlefield, 1990); Betty A. Reardon, *Sexism and the War System* (New York: Teachers College Press, 1985).

23. Consider the activities described and the language used in Jane Addams, *Peace and Bread in Time of War* (New York: King's Crown Press, 1945); Elise Boulding, *One Small Plot of Heaven* (Wallingford, PA: Pendle Hill, 1989); and Dorothy Day, *The Long Loneliness* (San Francisco: Harper & Row, 1952).

24. Pearl S. Buck, *The Exile* (New York: Triangle, 1936), 134–35.

25. Quoted in Sara Ruddick, *Maternal Thinking: Toward a Politics of Peace* (Boston: Beacon Press, 1989), 186. See also Adrienne Rich, *Of Woman Born* (New York: W. W. Norton, 1976).

26. Ruddick, *Maternal Thinking*, 189.

27. Quoted in Ruddick, *Maternal Thinking*, 154.

28. James, *Varieties of Religious Experience*, 359.

29. Michael S. Kimmel, "Clarence, William, Iron Mike, Tailhook, Senator Packwood, Spur Posse, Magic . . . and Us," in Buchwald, Fletcher, and Roth, *Transforming a Rape Culture*, 123.

30. Ibid.

31. Ibid., 122.

32. Christopher S. Queen, "The Peace Wheel: Nonviolent Activism in the Buddhist Tradition," in *Subverting Hatred: The Challenge of Nonviolence in Religious Traditions*, ed. Daniel L Smith-Christopher (Maryknoll, NY: Orbis Books, 1998), 25–47.

33. Jared Diamond, *Guns, Germs, and Steel* (New York: W. W. Norton, 2005), 278.

34. See the accounts in Jonathan Shay, *Achilles in Vietnam: Combat Trauma and the Undoing of Character* (New York: Scribner, 1994).

35. Ibid., 31.

36. Ibid.

37. Schrijvers, *GI War against Japan*, 208.

38. For more on these topics, see Noddings, *Challenge to Care in Schools* (New York: Teachers College Press, 2005); and Noddings, *Happiness and Education* (Cambridge: Cambridge University Press, 2003). See also Barbara Kingsolver, *Animal, Vegetable, Miracle* (New York: Harper Perennial, 2007).

CHAPTER 9. CONVERGENCE

1. See Marc D. Hauser, *Moral Minds* (New York: Harper Collins, 2006).

2. See Marc Bekoff, *Minding Animals* (Oxford: Oxford University Press, 2002).

3. See the argument in Susan Moller Okin, "Reason and Feeling in Thinking about Justice," *Ethics* 99, no. 2 (1989): 229–49.

4. Quoted in Sissela Bok, *Lying: Moral Choice in Public and Private Life* (New York: Vintage Books, 1979), 34.

5. For a well-documented account of the ethical lapses in the "war on terror," see Peter Jan Honigsberg, *Our Nation Unhinged: The Human Consequences of the War on Terror* (Berkeley and Los Angeles: University of California Press, 2009).

6. See Sara Ruddick, *Maternal Thinking: Toward a Politics of Peace* (Boston: Beacon Press, 1989).

7. Virginia Held, *The Ethics of Care: Personal, Political, and Global* (Oxford: Oxford University Press, 2006).

8. See Ruth Groenhout, *Connected Lives: Human Nature and an Ethics of Care* (Lanham, MD: Rowman & Littlefield, 2004).

BIBLIOGRAPHY

Addams, Jane. *Peace and Bread in Time of War.* New York: King's Crown Press, 1945/1915.

Ames, Roger T., Wimal Dissanayake, and Thomas P. Kasulis, eds. *Self as Person in Asian Theory and Practice.* Albany: State University of New York Press, 1994.

Babcock, Linda, and Sara Laschever. *Women Don't Ask: Negotiation and the Gender Divide.* Princeton, NJ: Princeton University Press, 2003.

Baier, Annette. *Moral Prejudices: Essays on Ethics.* Cambridge, MA: Harvard University Press, 1994.

Barnett, Rosalind, and Caryl Rivers. *Same Difference: How Gender Myths Are Hurting Our Relationships, Our Children, and Our Jobs.* New York: Basic Books, 2004.

Bekoff, Marc. *Minding Animals.* Oxford: Oxford University Press, 2002.

Bellamy, Edward. *Looking Backward.* New York: New American Library, 1960/1897.

Bentham, Jeremy. *Introduction to the Principles of Morals and Legislation.* Oxford: Clarendon Press, 1996/1789.

Berlin, Isaiah. *Four Essays on Liberty.* Oxford: Oxford University Press, 1969.

Bianchi, Suzanne M., Lynne M. Casper, and Rosalind Berkowitz King, eds. *Work, Family, Health, and Well-Being.* Mahwah, NJ: Lawrence Erlbaum, 2005.

Bloom, Paul. *Descartes' Baby.* New York: Basic Books, 2004.

Blum, Lawrence A. *Moral Perception and Particularity.* Cambridge: Cambridge University Press, 1994.

Bok, Sissela. *Lying: Choice in Public and Private Life.* New York: Vintage Books, 1979.

Boulding, Elise. *One Small Plot of Heaven.* Wallingford, PA: Pendle Hill, 1989.

Braybrooke, David. *Meeting Needs.* Princeton, NJ: Princeton University Press, 1987.

Brock-Utne, Birgit. *Educating for Peace: A Feminist Perspective.* New York: Pergamon Press, 1985.

Broom, Donald M. *The Evolution of Morality and Religion.* Cambridge: Cambridge University Press, 2003.

Buber, Martin. *I and Thou,* trans. Walter Kaufmann. New York: Charles Scribner's Sons, 1970.

Buchwald, Emilie, Pamela R. Fletcher, and Martha Roth, eds. *Transforming a Rape Culture.* Minneapolis: Milkweed Editions, 1993.

Buck, Pearl S. *The Exile.* New York: Triangle, 1936.

Butler, Samuel. *The Way of All Flesh.* Garden City, NY: Doubleday, 1944.

Cambridge Women's Peace Collective. *My Country Is the Whole World: An Anthology of Women's Work on Peace and War.* Boston: Pandora Press, 1984.

Candib, Lucy. *Medicine and the Family: A Feminist Perspective.* New York: Basic Books, 1995.

Card, Claudia. "Caring and Evil." *Hypatia* 5, no. 1 (1990): 101–8.

Carroll, Sean B. *The Making of the Fittest.* New York: W.W. Norton, 2006.

Chodorow, Nancy. *The Reproduction of Mothering.* Berkeley and Los Angeles: University of California Press, 1978.

Clement, Grace. *Care, Autonomy, and Justice: Feminism and the Ethic of Care*. Boulder, CO: Westview Press, 1996.

Cortright, David. *Peace: A History of Movements and Ideas*. Cambridge: Cambridge University Press, 2008.

Daly, Mary. *Beyond God the Father*. Boston: Beacon Press, 1974.

Damasio, Antonio. *Descartes' Error: Emotion, Reason, and the Human Brain*. New York: Putnam, 1994.

Dawkins, Richard. *The God Delusion*. Boston: Houghton Mifflin, 2006.

Day, Dorothy. *The Long Loneliness*. San Francisco: Harper & Row, 1952.

DeGrazia, David. *Taking Animals Seriously*. Cambridge: Cambridge University Press, 1996.

Dennett, Daniel C. *Breaking the Spell*. New York: Viking, 2006.

De Waal, Frans. *Primates and Philosophers: How Morality Evolved*. Princeton, NJ: Princeton University Press, 2006.

Dewey, John. *Experience and Education*. New York: Macmillan, 1963/1938.

Diamond, Jared. *Guns, Germs, and Steel*. New York: W.W. Norton, 2005.

Downing, Christine. *The Goddess*. New York: Crossroad, 1984.

Dugatkin, Lee Alan. *The Altruism Equation*. Princeton, NJ: Princeton University Press, 2006.

Early, Frances H. *A World without War*. Syracuse, NY: Syracuse University Press, 1997.

Ehrman, Bart D. *God's Problem*. New York: Harper Collins, 2008.

Eisler, Riane. *The Chalice and the Blade*. New York: Harper Collins, 1987.

———. *The Real Wealth of Nations*. San Francisco: Berrett-Koehler, 2007.

Elshtain, Jean Bethke. *Jane Addams and the Dream of American Democracy*. New York: Basic Books, 2002.

———. *Women and War*. New York: Basic Books, 1987.

Elshtain, Jean Bethke, and Sheila Tobias, eds. *Women, Militarism, and War*. Savage, MD: Rowman & Littlefield, 1990.

Entine, Jon. *Abraham's Children: Race, Identity, and the DNA of the Chosen People*. New York: Grand Central, 2007.

Filkins, Dexter. *The Forever War.* New York: Knopf, 2008.

Fineman, Martha. *The Autonomy Myth: A Theory of Dependency.* New York: New Press, 2004.

Foot, Philippa. *Natural Goodness.* Oxford: Oxford University Press, 2001.

Frankl, Viktor E. *The Doctor and the Soul.* New York: Vintage Books, 1973.

Gardner, Martin. *The Whys of a Philosophical Scrivener.* New York: Quill, 1983.

Gilligan, Carol J. *In a Different Voice.* Cambridge, MA: Harvard University Press, 1982.

Gilligan, James. *Violence.* New York: G. P. Putnam's Sons, 1996.

Glover, Jonathan. *Humanity: A Moral History of the 20th Century.* New Haven, CT: Yale University Press, 2000.

Gray, J. G. *The Warriors: Reflections on Men in Battle.* Lincoln, NE: Bison Press, 1998/1959.

Gray, John, *Black Mass: Apocalyptic Religion and the Death of Utopia.* Canada: Doubleday, 2007.

Groenhout, Ruth. *Connected Lives: Human Nature and an Ethics of Care.* Lanham, MD: Rowman & Littlefield, 2004.

Halevi, Yossi Klein. *At the Entrance to the Garden of Eden.* New York: Perennial, 2001.

Halwani, Raja. "Care Ethics and Virtue Ethics." *Hypatia* 18, no. 3 (2003): 161–92.

Hauser, Marc D. *Moral Minds.* New York: Harper Collins, 2006.

Hedges, Chris. "War Is a Force That Gives Us Meaning." *Amnesty Now* (Winter 2002): 10–13.

Held, Virginia. *The Ethics of Care: Personal, Political, and Global.* Oxford: Oxford University Press, 2006.

———. "Moral Subjects: The Natural and the Normative." *Proceedings and Addresses of the American Philosophical Association* 76, no. 2 (2002): 7–24.

Hirshman, Linda. *Get to Work . . . and Get a Life, Before It's Too Late.* New York: Penguin, 2007.

Hoagland, Sarah Lucia. "Some Concerns about Nel Noddings' *Caring*." *Hypatia* 5, no. 1 (1990): 109–14.

Hoffman, Martin. *Empathy and Moral Development: Implications for Caring and Justice*. New York: Cambridge University Press, 2000.

Honigsberg, Peter Jan. *Our Nation Unhinged: The Human Consequences of the War on Terror.* Berkeley and Los Angeles: University of California Press, 2009.

Houston, Barbara. "Caring and Exploitation." *Hypatia* 5, no. 1 (1990): 115–19.

Hrdy, Sarah Blaffer. *Mothers and Others.* Cambridge, MA: Belknap Press of Harvard University Press, 2009.

Hume, David. *An Enquiry Concerning the Principles of Morals.* Indianapolis: Hackett, 1983/1751.

Ickes, William W. *Empathic Accuracy.* New York: Guilford, 1997.

James, William. *The Varieties of Religious Experience.* New York: Modern Library, 1929/1902.

Joyce, Richard. *The Evolution of Morality.* Cambridge, MA: MIT Press, 2006.

Katz, Michael S., Nel Noddings, and Kenneth A. Strike, eds. *Justice and Caring: The Search for Common Ground in Education.* New York: Teachers College Press, 1999.

Kimball, Charles. *When Religion Becomes Evil.* New York: Harper Collins, 2002.

Kimmel, Michael S. "Clarence, William, Iron Mike, Tailhook, Senator Packwood, Spur Posse, Magic . . . and Us." In *Transforming a Rape Culture,* ed. Emilie Buchwald, Pamela R. Fletcher, and Martha Roth, 121–38. Minneapolis: Milkweed Editions, 1993.

Kingsolver, Barbara. *Animal, Vegetable, Miracle.* New York: Harper Perennial, 2007.

Kittay, Eva Feder. *Love's Labor: Essays on Women, Equality, and Dependency.* New York: Routledge, 1999.

Kohlberg, Lawrence. *The Philosophy of Moral Development.* Vol. 1. San Francisco: Harper & Row, 1981.

Levi, Primo. *The Drowned and the Saved*, trans. Raymond Rosenthal. New York: Vintage, 1988.

Levitin, Daniel J. *The World in Six Songs*. New York: Dutton, 2008.

Li, Chenyang. "The Confucian Concept of Jen and the Feminist Ethics of Care: A Comparative Study." *Hypatia* 9, no. 1 (1994): 70–89.

Lindbergh, Anne Morrow. *Gift from the Sea*. New York: Random House, 1955.

Linden, David L. *The Accidental Mind*. Cambridge, MA: Belknap Press of Harvard University Press, 2007.

MacIntyre, Alasdair. *After Virtue*. Notre Dame, IN: University of Notre Dame Press, 1981.

———. *Whose Justice? Which Rationality?* Notre Dame, IN: Notre Dame University Press, 1988.

Mackenzie, Catriona, and Natalie Stoljar, eds. *Relational Autonomy*. Oxford: Oxford University Press, 2000.

MacKinnon, Catherine A. *Feminism Unmodified*. Cambridge, MA: Harvard University Press, 1987.

Mahowald, Mary Briody, ed. *Philosophy of Woman*. Indianapolis: Hackett, 1983.

Martin, Jane Roland. *Reclaiming a Conversation*. New Haven, CT: Yale University Press, 1985.

Maslow, Abraham H. *The Farther Reaches of Human Nature*. New York: Viking Press, 1971.

———. *Motivation and Personality*. New York: Harper & Row, 1970.

Masson, Jeffrey Moussaieff, and Susan McCarthy. *When Elephants Weep: The Emotional Lives of Animals*. New York: Delacorte Press, 1995.

Meyers, Diana T. *Self, Society, and Personal Choice*. New York: Columbia University Press, 1989.

Michel, Sonya. *Children's Interests/Mothers' Rights*. New Haven, CT: Yale University Press, 1999.

Murdoch, Iris. *The Sovereignty of Good*. London: Routledge & Kegan Paul, 1970.

Noddings, Nel. *Caring: A Feminine Approach to Ethics and Moral Education.* Berkeley and Los Angeles: University of California Press, 1984.

———. *The Challenge to Care in Schools.* New York: Teachers College Press, 1992.

———. *Critical Lessons: What Our Schools Should Teach.* Cambridge: Cambridge University Press, 2006.

———. *Educating for Intelligent Belief or Unbelief.* New York: Teachers College Press, 1993.

———. *Educating Moral People.* New York: Teachers College Press, 2002.

———. *Happiness and Education.* Cambridge: Cambridge University Press, 2003.

———. *Starting at Home: Caring and Social Policy.* Berkeley and Los Angeles: University of California Press, 2002.

———. *Women and Evil.* Berkeley and Los Angeles: University of California Press, 1989.

Nussbaum, Martha C. *Sex and Social Justice.* New York: Oxford University Press, 1999.

Okin, Susan Moller. "Reason and Feeling in Thinking about Justice." *Ethics* 99, no. 2 (1989): 229–49.

Oliner, Samuel P., and Pearl M. Oliner. *The Altruistic Personality: Rescuers of Jews in Nazi Europe.* New York: The Free Press, 1988.

Orwell, George. *Nineteen Eighty-Four.* New York: Harcourt, Brace, & World, 1949.

Phillips, John Anthony. *Eve: The History of an Idea.* San Francisco: Harper & Row, 1984.

Piaget, Jean. *Biology and Knowledge.* Chicago: University of Chicago Press, 1971.

Pogge, Thomas. *World Poverty and Human Rights.* Cambridge: Polity/Blackwell, 2002.

Queen, Christopher S. "The Peace Wheel: Nonviolent Activism in the Buddhist Tradition." In *Subverting Hatred: The Challenge of Nonviolence in Religious Traditions,* ed. Daniel L. Smith-Christopher, 25–47. Maryknoll, NY: Orbis Books, 1998.

Rawls, John. *A Theory of Justice.* Cambridge, MA: Harvard University Press, 1971.

Reardon, Betty A. *Sexism and the War System.* New York: Teachers College Press, 1985.

Reverby, Susan. *Ordered to Care.* Cambridge: Cambridge University Press, 1987.

Rhode, Deborah L., ed. *Theoretical Perspectives on Sexual Difference.* New Haven, CT: Yale University Press, 1990.

Rich, Adrienne. *Of Woman Born.* New York: W. W. Norton, 1976.

Ricoeur, Paul. *The Symbolism of Evil.* Boston: Beacon Press, 1969.

Robinson, Fiona. *Globalizing Care.* Boulder, CO: Westview Press, 1999.

Ruddick, Sara. *Maternal Thinking: Toward a Politics of Peace.* Boston: Beacon Press, 1989.

Sandel, Michael. *Liberalism and the Limits of Justice.* Cambridge: Cambridge University Press, 1982.

Sartre, Jean-Paul. *Being and Nothingness,* trans. Hazel E. Barnes. New York: Washington Square Press, 1956.

———. *Nausea,* trans. Lloyd Alexander. Norfolk, CT: New Directions, 1959.

Saunders, Harold H. *The Other Walls.* Princeton, NJ: Princeton University Press, 1991.

Schrijvers, Peter. *The GI War against Japan.* New York: New York University Press, 2002.

Shay, Jonathan. *Achilles in Vietnam: Combat Trauma and the Undoing of Character.* New York: Scribner, 1994.

Singer, Peter. *One World: The Ethics of Globalization.* New Haven, CT: Yale University Press, 2002.

Slote, Michael. "Caring in the Balance." In *Norms and Values: Essays on the Work of Virginia Held,* ed. Joram G. Haber and Mark S. Halfon, 27–36. Lanham, MD: Rowman & Littlefield, 1998.

———. *The Ethics of Care and Empathy.* New York: Routledge, 2007.

———. "Famine, Affluence, and Virtue." In *Working Virtue,* ed. Rebecca L. Walker and Philip J. Ivanhoe, 279–96. Oxford: Oxford University Press, 2007.

————. *Moral Sentimentalism*. Oxford: Oxford University Press, 2010.

Smail, Daniel Lord. *On Deep History and the Brain*. Berkeley and Los Angeles: University of California Press, 2007.

Smith, Christian. *Soul Searching: The Religious and Spiritual Lives of American Teenagers*. Oxford: Oxford University Press, 2005.

Star, Daniel. "Do Confucians Really Care? A Defense of the Distinctiveness of Care." *Hypatia* 17, no. 1 (2002): 77–106.

Stiglitz, Joseph E. *Globalization and Its Discontents*. New York: W.W. Norton, 2002.

Stocker, Michael. *Valuing Emotions*. Cambridge: Cambridge University Press, 1996.

Stone, Merlin. *When God Was a Woman*. New York: Dial Press, 1976.

Stueber, Karsten R. *Rediscovering Empathy*. Cambridge, MA: MIT Press, 2006.

Swofford, Anthony. *Jarhead*. New York: Scribner, 2003.

Taylor, Charles. *Sources of the Self*. Cambridge, MA: Harvard University Press, 1989.

Tierney, Kevin. *Darrow: A Biography*. New York: Thomas Y. Crowell, 1979.

Tillich, Paul. *The Courage to Be*. New Haven, CT: Yale University Press, 1952.

Tronto, Joan. *Moral Boundaries: A Political Argument for an Ethic of Care*. New York: Routledge, 1993.

True, Michael. *An Energy Field More Intense Than War*. Syracuse, NY: Syracuse University Press, 1995.

Turner, Jonathan H. *On the Origins of Human Emotions*. Stanford, CA: Stanford University Press, 2000.

Unger, Peter. *Living High and Letting Die*. New York: Oxford University Press, 1996.

Verducci, Susan. "A Conceptual History of Empathy and a Question It Raises for Moral Education." *Educational Theory* 50, no. 1 (2001): 63–80.

Walker, Margaret Urban. "Moral Understandings: Alternative 'Epistemology' for a Feminist Ethics." *Hypatia* 4 (Summer 1989): 15–28.

Walker, Rebecca L., and Philip J. Ivanhoe, eds. *Working Virtue*. Oxford: Oxford University Press, 2007.

Walzer, Michael. *Just and Unjust Wars*. New York: Basic Books, 1977.

Watson, Jean. *Nursing: Human Science and Human Care*. Norwalk, CT: Appleton-Century-Crofts, 1985.

Weil, Simone. *Simone Weil Reader*, ed. George A. Panichas. Mt. Kisco, NY: Moyer Bell Limited, 1977.

White, James Terry. *Character Lessons in American Biography*. New York: Character Development League, 1909.

Wiesenthal, Simon. *The Sunflower*. New York: Schocken Books, 1976.

Wilson, Edward O. *The Creation: An Appeal to Save Life on Earth*. New York: W. W. Norton, 2006.

Wissler, Clark. *Indians of the United States*. New York: Anchor Books, 1989.

Woolf, Virginia. *Collected Essays*. Vol. 2. London: Hogarth Press, 1966.

Wrangham, Richard, and Dale Peterson. *Demonic Males: Apes and the Origins of Human Violence*. New York: Houghton Mifflin, 1996.

Yoder, John Howard. *The Politics of Jesus*. Grand Rapids, MI: Eerdmans, 1994.

Zimmerman, Mary K., Jacqueline S. Litt, and Christine E. Bose, eds. *Global Dimensions of Gender and Carework*. Stanford, CA: Stanford University Press, 2006.

INDEX

acceptability, need for, 73, 191, 196–203
Addams, Jane, 222
Adorno, Theodore, 166
Advanced Placement courses, 122, 201
Afghanistan, war in, 247
aggression, male, 4, 172–73, 248. *See also* violence; war; culture and socialization supporting, 215, 225; directed at women, 225; evolutionary tendency to, 4, 8–9, 205, 220, 222
altruism, 1, 2; distinction between egoism and, 136–37; in nonhuman species, 14; reciprocal, 41
"America the Beautiful," 223–24
Angel in the House, myth of, 153
anger, 31, 65, 172
animals, nonhuman, 235, 240; caring for, as pets, 61–62; emotions in, 161, 164; parental behavior in, 10–11, 27; social instincts of, 165, 170

Apostle's Creed, 149
Aquinas, Thomas, 173
Aristotelianism, 24, 175
Aristotle, 32
attention, 28, 36, 162–63, 243; to acceptability needs, 191, 202; to assumed needs, 59–60; in caring relation, 118, 130; conscious, of mothers, 34; empathy and, 51–52, 56, 63–64, 73, 77; to expressed needs, 31, 59, 192, 203; proximity and, 51; receptive, 7, 12, 47–48, 52–53, 152, 174; sympathetic, 56, 70, 137, 163, 174–76
Augustine, 141, 142, 144, 150, 206
Austen, Jane, 161
authority: moral, 237–38; obedience to, 63, 123, 146, 227; religious, 139, 142, 144–47, 151, 155, 226, 241, 245; traditional, of communitarianism, 127; in war, 206, 227
authority figures, 131, 193

Text: 10/15 Janson

Display: Janson

Compositor: Toppan Best-set Premedia Limited

Indexer: Ruth Elwell

Printer and binder: Maple-Vail Book Manufacturing Group